HACCP

Establishing Hazard Analysis

Critical Control Point Programs

A WORKSHOP MANUAL

edited by

Kenneth E. Stevenson, Ph.D.

Principal Research Scientist

and

Dane T. Bernard

*Vice President,
Food Safety and Strategic Programs*

NATIONAL FOOD PROCESSORS ASSOCIATION

published by

The Food Processors Institute

1401 New York Avenue, N.W.
Washington, D.C. 20005

SECOND EDITION, 7th Printing
© 1995

Training 4/99

DW

SECOND EDITION, 7th Printing

© 1995 The Food Processors Institute
Washington, D.C.

LIBRARY OF CONGRESS
CATALOG CARD NO.: 95-060297

Establishing Hazard Analysis Critical Control Point Programs
A Workshop Manual

Washington, D.C.: Food Processors Institute, The

224 p.

ISBN 0-937774-03-0

Foreword to the Second Edition

Since publication of the first edition of *HACCP - Establishing Hazard Analysis Critical Control Point Programs: A Workshop Manual*, we have trained thousands of individuals in workshops on the principles and applications of HACCP. In addition, we have worked with individual companies to develop and implement their own HACCP plans. Recently, significant initiatives by both the U.S. Food and Drug Administration and the U.S. Department of Agriculture have resulted in proposed regulations which will doubtless shape our future food safety policy and the HACCP programs developed in response to the regulatory initiatives. In this second edition of our HACCP manual, we have incorporated changes in an attempt to capture the current understanding of the following:

- Hazard analysis;

- Clarification of chemical and physical hazards; and

- The importance of prerequisite programs including hygienic practices and employee training.

This manual was developed as a basic text covering the seven principles of HACCP as defined by the National Advisory Committee on Microbiological Criteria for Foods. The text is non-product category specific and utilized by all segments of the food industry as a training vehicle. We hope you will find the second edition as useful as those who have used the first edition.

In addition to those listed as chapter authors, the editors thank Austin Gavin, Michael Jantschke, Lisa Nesbett, and Lisa Weddig for their contributions and assistance in providing HACCP training. We also thank Rita Fullem and Susan Gray of The Food Processors Institute for their assistance in course organization and coordination.

LIST OF CONTRIBUTORS

Dane T. Bernard
Vice President - Food Safety
and Strategic Programs
National Food Processors Assoc.
1401 New York Avenue, NW
Washington, D.C. 20005

Cleve B. Denny
Consultant
6230 Valley Road
Bethesda, MD 20817

Lloyd R. Hontz
Director (USDA) - Technical
Regulatory Affairs
National Food Processors Assoc.
1401 New York Avenue, NW
Washington, D.C. 20005

John Y. Humber
Research Principal
Kraft Foods, Inc.
801 Waukegan Road
Glenview, IL 60025

Bonnie J. Humm
Consultant
2019 Fuller Street
Mountain Home, AR 72653

Allen M. Katsuyama
Manager - Quality Assurance
Center for Technical Services and Assistance
National Food Processors Assoc.
6363 Clark Avenue
Dublin, CA 94568

Lloyd J. Moberg
Vice President - Quality
Borden Foods International
180 E. Broad Street
Columbus, OH 43215

Virginia N. Scott
Director - Processing Technology
and Microbiology Group
National Food Processors Assoc.
1401 New York Avenue, NW
Washington, D.C.

Kenneth E. Stevenson
Principal Research Scientist - Center for
Technical Services and Assistance
National Food Processors Assoc.
6363 Clark Avenue
Dublin, CA 94568

TABLE OF CONTENTS

SECTION I - HACCP AND HACCP PRINCIPLES

Chapter 1 - Introduction to HACCP Systems

Chapter 2 - Hazard Analysis and Critical Control Point System

Chapter 3 - Basic Steps in The Development of HACCP Systems

SECTION II - HAZARDS AND CONTROLS

Chapter 4 - Biological Hazards and Controls

Chapter 5 - Chemical Hazards & Controls

Chapter 6 - Physical Hazards and Controls

SECTION III - DEVELOPING HACCP PLANS

Chapter 7 - Initial Tactics in Developing HACCP Plans

Chapter 8 - Hazard Analysis and Identification of CCPs

Chapter 9 - CLs, Monitoring and Corrective Actions

Chapter 10 - Recordkeeping and Verification Procedures

Chapter 11 - Workshop Flow Diagrams and Forms

SECTION IV - MANAGING HACCP PROGRAMS

SECTION V - HACCP AND REGULATIONS

SECTION I

HACCP AND HACCP PRINCIPLES

Introduction to Hazard Analysis Critical Control Point Systems

Hazard Analysis and Critical Control Point System

Basic Steps in the Development of HACCP Systems

CHAPTER 1

INTRODUCTION TO HAZARD ANALYSIS CRITICAL CONTROL POINT SYSTEMS

by K. E. Stevenson

HACCP CONCEPT

The **H**azard **A**nalysis **C**ritical **C**ontrol **P**oint (HACCP) system is a preventive system for assuring the safe production of food products. It is based on a common-sense application of technical and scientific principles to the food production process from field to table. The principles of HACCP are applicable to all phases of food production, including basic agriculture, food preparation and handling, food processing, food service, distribution systems and consumer handling and use.

The most basic concept underlying HACCP is that of prevention rather than inspection. A food grower, processor, handler, distributor or consumer should have sufficient information concerning the food and the related procedures they are using, so they will be able to identify where a food safety problem may occur and how it will occur. If the "where" and "how" are known, prevention becomes easy and obvious, and finished product inspection and testing becomes superfluous. A HACCP program deals with control of factors affecting the ingredients, product and process. The objective is to make the product safely, <u>and</u> to be able to prove that the product has been made safely. The where and how are the *HA* (Hazard Analysis) part of HACCP. The proof of the control of processes and conditions is the *CCP* (Critical Control Point) part. Flowing from this basic concept, HACCP is simply a methodical and systematic application of the appropriate science and technology to plan, control and document the safe production of foods.

The practical definition of HACCP should clearly state that the HACCP concept covers all types of potential food safety hazards--biological, physical, and chemical--whether they are naturally occurring in the food, contributed by the environment or generated by a mistake in the manufacturing process. While chemical hazards are the most feared by consumers and physical hazards are the most commonly identified by consumers, microbiological hazards are the most serious from a public health perspective. For this reason, while HACCP systems address all 3 types of hazards, a majority of the emphasis is placed on microbiological issues. For example, a piece of metal (physical hazard) in a food product may result in a chipped tooth for one consumer, but contamination of a batch of milk with *Salmonella* may affect hundreds or even thousands of consumers.

ORIGIN OF HACCP

Development of Foods for the Space Program

The HACCP system was developed by the Pillsbury Company in response to the food safety requirements imposed by NASA for "space foods" produced for manned space flights beginning in 1959. NASA had two principle safety issues. The first was related to potential problems with food particles--crumbs--in the space capsule under conditions of zero gravity. (They were concerned about potential problems of crumbs interfering with electrical equipment.) The second issue was the absolute assurance of freedom from pathogens and biological toxins. A case of foodborne illness, e.g. staphylococcal food poisoning, in a space capsule, would have been catastrophic.

The first concern, food crumbs in zero gravity, was addressed by developing bite-sized foods and by using specially formulated edible coatings to hold the food together. In addition, highly specialized types of packaging were used to minimize the exposure of foods to the environment during storage, preparation and consumption. The second concern, microbiological safety was more difficult to address. Sampling of finished product, to establish microbiological safety of each batch of space food produced, proved to be impractical, if not impossible. To quote Dr. Howard Bauman, who managed the development of HACCP at Pillsbury:

> "We quickly found that by using standard methods of quality control there was absolutely no way we could be assured that there wouldn't be a problem. This brought into serious question the then prevailing system of quality control in our plants... If we had to do a great deal of destructive testing to come to a reasonable conclusion that the product was safe to eat, how much were we missing in the way of safety issues by principally testing only the end product and raw materials?
>
> We concluded after extensive evaluation that the only way we could succeed would be to establish control over the entire process, the raw materials, the processing environment and the people involved."

To help quantify the impracticality of attribute sampling and the resultant destructive testing of end product which would be necessary to assure microbiological safety, consider the following: If *Salmonella* was present in a batch of product at the rate of 1 out of every 1000 units of product (defect rate = 0.1%), a sampling plan which analyzed 60 units from the batch would have >94% probability of approving the batch--and missing the salmonella-contaminated product.

In addition to the statistical evidence that this sampling plan would be ineffective in detecting the contaminated product, there is the practical and economic reality that no company would be able to afford to destructively test 60 units out of every batch of product for the presence of *Salmonella*. Thus, an alternative approach had to be developed in order to obtain the level of assurance of product safety which NASA required for foods produced for the space program.

At first, they explored the use of NASA's "Zero Defects Program" which was designed for testing hardware intended for the space program. This program utilized a series of non-destructive tests of hardware for the purpose of assuring that the hardware functioned properly. While repeated, non-destructive testing could be used on every piece of hardware, this program was not appropriate for adaptation to foods.

Eventually, the Modes of Failure concept developed by the U.S. Army Natick Laboratories, was adapted to the production of foods. By gathering knowledge and experience concerning a food product/process, it was possible to predict what might go wrong (a "hazard"), how it would occur, and where it would occur in the process. Based on this type of analysis of the hazards associated with a specific product and process, it was possible to select points at which measurements and/or observations could be made which would demonstrate whether or not the process was being controlled. If the process was out of control, there was an increased probability that a food safety problem would occur. These points in the process were then, and are today, called Critical Control Points (CCPs). Thus, HACCP was developed to target proper design of all of the factors associated with ingredients, processes and products in order to prevent hazards from occurring, and thereby ensure the safety of the products.

The Original HACCP System

The HACCP concept was first presented to the public at the 1971 National Conference on Food Protection. This initial HACCP system consisted of three principles:

1. Identification and assessment of hazards associated with growing/harvesting to marketing/preparation.

2. Determination of the critical control points to control any identifiable hazard.

3. Establishment of systems to monitor critical control points.

Along with these principles, the system identified a CCP as a point in the manufacture of a product whose loss of control would result in an unacceptable food safety risk.

The preventive nature of the HACCP system is readily apparent when these principles are paraphrased, as follows:

- Identify any safety-related problems which are related to this product and process.
- Determine the specific factors which need to be controlled to prevent these problems from occurring.
- Establish systems that can measure and document whether or not these factors are being controlled properly.

EARLY USES OF HACCP

At first, there was considerable interest in this new approach to food safety. The U.S. Food and Drug Administration (FDA) began training its inspectors in the elements of HACCP, and they instituted special HACCP inspections of food plants. There were numerous conferences and sessions on HACCP, including a symposium at the 1974 Annual Meeting of the Institute of Food Technologists.

During the 1970's, FDA promulgated the low-acid and acidified canned food regulations--Title 21, Code of Federal Regulations Part 113 (originally 21 CFR 128b), "Thermally Processed Low-Acid Foods Packaged in Hermetically Sealed Containers," and 21 CFR 114, "Acidified Foods". While these regulations did not mention HACCP, they certainly appear to be based upon HACCP concepts.

After this initial flurry of activity, interest in HACCP appeared to wane. While the description of the HACCP principles is relatively brief, developing a HACCP program is not a simple matter. It takes considerable time and expertise to set up a HACCP program. Therefore, except for use by a few large food companies and the required use of HACCP concepts for FDA-regulated thermally processed low-acid and acidified foods, HACCP has not been widely used in the food industry.

1985 NAS REPORT

Interest in HACCP was rekindled in 1985 when a Subcommittee of the Food Protection Committee of the National Academy of Sciences (NAS) issued a report on microbiological criteria. This report was the result of a study commissioned by several government agencies with responsibilities for food safety. While the objectives of the study were mainly related to establishing microbiological criteria for foods, the report included a particularly strong endorsement of HACCP. The report recommended that regulators and industry both utilize HACCP because it was the most effective and efficient means of assuring the safety of our food supply.

While the 1985 NAS report received mostly favorable responses, two areas have elicited some unfavorable responses:

1. The statement that HACCP would have to be required by regulation if it is to be widely utilized.

2. The apparent approval of regulatory access to a variety of records.

NATIONAL ADVISORY COMMITTEE ON MICROBIOLOGICAL CRITERIA FOR FOODS

Based upon recommendations in the 1985 NAS report, a committee, consisting primarily of food microbiologists, was appointed to serve as an expert scientific advisory panel to the Secretaries of Agriculture, Commerce, Defense and Health and Human Services.

This committee held its first meetings in 1988, and was named the National Advisory Committee on Microbiological Criteria for Foods (NACMCF). Part of the mission of the NACMCF is to encourage adoption of the HACCP approach to food safety. In their initial meetings and discussions, it became obvious that there were several different opinions concerning the specifics associated with HACCP systems. Therefore, a HACCP group was appointed to study HACCP and make recommendations to NACMCF.

The chapters which follow provide details concerning the HACCP Principles which were developed by the NACMCF, and additional information for use in applying the HACCP Principles to food processing operations.

REFERENCES

DHEW. 1971. *Proceedings of the 1971 National Conference on Food Protection*. U.S. Department of Helath, Education and Welfare, Public Health Service, Washington, D.C.

NAS. 1985. *An Evaluation of the Role of Microbiological Criteria for Foods and Food Ingredients*. National Academy Press, Washington, D.C.

Pillsbury Company. 1973. *Food Safety Through the Hazard Analysis Critical Control Point System*. Contract No. FDA 72-59. Research and Development Dept., The Pillsbury Company, Minneapolis, MN.

FDA. 1992. Acidified foods. Title 21, *Code of Federal Regulations*, Part 114. U.S. Government Printing Office, Washington, D.C.

FDA. 1992. Thermally processed low-acid foods packaged in hermetically sealed containers. Title 21, *Code of Federal Regulations*. U.S. Government Printing Office, Washington, D.C.

CHAPTER 2

HAZARD ANALYSIS AND CRITICAL CONTROL POINT SYSTEM

by National Advisory Committee On Microbiological Criteria For Foods

EDITORIAL NOTE

In 1989, an ad hoc working group of the National Advisory Committee on Microbiological Criteria for Foods (NACMCF) was formed to establish guidelines for application of HACCP. Based upon the material developed by the working group, the NACMCF adopted a document entitled, "HACCP Principles for Food Production," in November, 1989. In that document, the NACMCF defined HACCP as "a systematic approach to be used in food production as a means to assure food safety," endorsed the use of HACCP by industry and regulators, described seven HACCP principles, and provided a "guide for HACCP Plan development for a specific food."

The seven HACCP principles were scrutinized by industry and regulatory personnel, and, in general, the concepts were favorably accepted. However, in various meetings, courses and workshops, the usefulness of the microbiological risk assessment -- included as part of the description of the application of Principle No. 1 -- was questioned. The risk assessment consisted of ranking a food according to six general hazard characteristics and then assigning a risk category. Problems in employing the risk assessment included differences in interpretation of the general hazard characteristics and the fact that there was no direct link between Principle No. 1 (hazard analysis) and Principle No. 2 (determination of CCPs).

Subsequently, the Codex Food Hygiene Committee HACCP Working Group drafted a report on the use of HACCP which included a slightly different approach to the application of (slightly revised) HACCP principles. This included a hazard analysis, to identify hazards and preventive measures (Principle No. 1), and application of a series of questions, named a "HACCP decision tree," to determine CCPs (Principle No. 2).

In 1991, the NACMCF reconvened the HACCP Working Group to review the NACMCF November, 1989 report. The Working Group drafted a new document which included modifications to the seven HACCP principles. The most significant modifications were made to Principles 1 and 2, and they appeared to be patterned after those of the Codex draft report. The NACMCF adopted the new document, "Hazard Analysis and Critical Control Point System," on March 20, 1992, and it is presented in its entirety in the remaining pages of this Chapter.

(Minor editorial changes have been made in format.)

EXECUTIVE SUMMARY

The National Advisory Committee on Microbiological Criteria for Foods (Committee) reconvened a Hazard Analysis Critical Control Point (HACCP) Working Group in July 1991. The primary purpose of the working group was to review the committee's November 1989 HACCP document comparing it with a draft report prepared by a HACCP Working Group of the Codex Committee on Food Hygiene. Based upon its review, the Committee has determined to expand upon its initial report by emphasizing the concept of prevention, incorporating a decision tree intended to facilitate the identification of Critical Control Points (CCPs), and providing a more detailed explanation of the application of HACCP principles.

The Committee again endorses HACCP as an effective and rational means of assuring food safety from harvest to consumption. Preventing problems from occurring is the paramount goal underlying any HACCP system. Seven basic principles are employed in the development of HACCP plans that meet the stated goal. These principles include hazard assessment, CCP identification, establishing critical limits, monitoring procedures, corrective actions, documentation, and verification procedures. Under such systems, if a deviation occurs indicating that control has been lost, appropriate steps are taken to reestablish control in a timely manner to assure that potentially hazardous products do not reach the consumer.

In the application of HACCP, the use of microbiological testing is seldom an effective means of monitoring critical control points (CCPs) because of the time required to obtain results. In most instances, monitoring of CCPs can best be accomplished through the use of physical and chemical tests, and through visual observations. Microbiological criteria do, however, play a role in verifying that the overall HACCP system is working.

The Committee believes that the HACCP principles should be standardized to create uniformity in its work, and in training and applying the HACCP system by industry and regulatory authorities. In accordance with the National Academy of Sciences recommendation, the HACCP system must be developed by each food establishment and tailored to its individual products, processing and distribution conditions.

In keeping with its charge of providing recommendations to its sponsoring agencies regarding microbiological food safety issues, this document focuses on this area. The Committee recognizes that in order to assure food safety, properly designed HACCP plans must also consider chemical and physical hazards in addition to other biological hazards.

In order for a successful HACCP program to be implemented, management must be committed to a HACCP approach. A commitment by management will indicate an awareness of the benefits and costs of HACCP and include education and training of employees. Benefits, in addition to food safety, are better use of resources and timely response to problems.

The Committee designed this document to guide the food industry in the implementation of HACCP systems. The Committee recommends that future documents address the role of regulatory agencies in the HACCP system.

DEFINITIONS

CCP Decision tree: A sequence of questions to determine whether a control point is a CCP.

Continuous Monitoring: Uninterrupted collection and recording of data such as temperature on a strip chart.

Control: (a) To manage the conditions of an operation to maintain compliance with established criteria. (b) The state wherein correct procedures are being followed and criteria are being met.

Control Point: Any point, step, or procedure at which biological, physical, or chemical factors can be controlled.

Corrective Action: Procedures to be followed when a deviation occurs.

Criterion: A requirement on which a judgment or decision can be based.

Critical Control Point (CCP): A point, step, or procedure at which control can be applied and a food safety hazard can be prevented, eliminated, or reduced to acceptable levels.

Critical Defect: A deviation at a CCP which may result in a hazard.

Critical Limit: A criterion that must be met for each preventive measure associated with a critical control point.

Deviation: Failure to meet a critical limit.

HACCP Plan: The written document which is based upon the principles of HACCP and which delineates the procedures to be followed to assure the control of a specific process or procedure.

HACCP System: The result of the implementation of the HACCP plan.

HACCP Team: The group of people who are responsible for developing a HACCP plan.

HACCP Plan Reevaluation: One aspect of verification in which a documented periodic review of the HACCP plan is done by the HACCP team with the purpose of modifying the HACCP plan as necessary.

HACCP Plan Validation: The initial review by the HACCP team to ensure that all elements of the HACCP plan are accurate.

Hazard: A biological, chemical, or physical property that may cause a food to be unsafe for consumption.

Monitor: To conduct a planned sequence of observations or measurements to assess whether a CCP is under control and to produce an accurate record for future use in verification.

Preventive Measure: Physical, chemical, or other factors that can be used to control an identified health hazard.

Random Checks: Observations or measurements which are performed to supplement the scheduled evaluations required by the HACCP plan.

Risk: An estimate of the likely occurrence of a hazard.

Sensitive Ingredient: An ingredient known to have been associated with a hazard and for which there is reason for concern.

Severity: The seriousness of a hazard.

Target Levels: Criteria which are more stringent than critical limits and which are used by an operator to reduce the risk of a deviation.

Verification: The use of methods, procedures, or tests in addition to those used in monitoring to determine if the HACCP system is in compliance with the HACCP plan and/or whether the HACCP plan needs modification and revalidation.

PURPOSE AND PRINCIPLES

HACCP is a systematic approach to food safety consisting of seven principles:

1. Conduct a hazard analysis. Prepare a list of steps in the process where significant hazards occur and describe the preventive measures.

2. Identify the CCPs in the process.

3. Establish critical limits for preventive measures associated with each identified CCP.

4. Establish CCP monitoring requirements. Establish procedures for using the results of monitoring to adjust the process and maintain control.

5. Establish corrective actions to be taken when monitoring indicates that there is a deviation from an established critical limit.

6. Establish effective record-keeping procedures that document the HACCP system.

7. Establish procedures for verification that the HACCP system is working correctly.

EXPLANATION AND APPLICATION OF PRINCIPLES

The HACCP concept is relevant to all stages throughout the food chain from growing, harvesting, processing, manufacturing, distributing, and merchandising to preparing food for consumption. Certain points in the food chain are better suited to the application of the HACCP principles. For example, food manufacturing facilities are very well suited to the adoption of the HACCP concept. The Committee recommends the adoption of HACCP to the fullest extent possible and reasonable throughout the food chain.

The Committee recognizes that education and training is an important element of the HACCP concept. Employees who will be responsible for the HACCP program must be adequately trained in the principles of HACCP, its application and implementation. However, education and training programs do not have to be limited to those directly involved with HACCP and its implementation. Education and training programs should be designed to address the needs of industry, government and academic personnel, as well as consumers. Educating home food handlers in the recognition and application of critical control points will improve the safety of food prepared in the home. It is recommended that educational materials be provided to home food handlers that address the safe acquisition and proper handling of foods.

Figure 2-1 lists steps used in the application of Principle 1.

Figure 2-1. First Six Steps for the Development of a HACCP Plan

1.	Assemble the HACCP Team

2.	Describe the Food and its Distribution

3.	Identify Intended Use and Consumers of the Food

4.	Develop Flow Diagram

5.	Verify Flow Diagram

6.	**Conduct Hazard Analysis** (a) Identify and List Steps in the Process Where the Hazards of Potential Significance Occur. (b) List All Identified Hazards Associated with Each Step. (c) List Preventive Measures to Control Hazards.

Step	Identified Hazard	Preventive Measures

ASSEMBLE THE HACCP TEAM

The first step in developing a HACCP plan is to assemble a HACCP team consisting of individuals who have specific knowledge and expertise appropriate to the product and process. It is the team's responsibility to develop each step of the HACCP plan. The team should be multi-disciplinary (e.g., engineering, production, sanitation, quality assurance, food microbiology). The team should include local personnel who are directly involved in the daily processing activities as they are more familiar with the variability and limitations of the operation. In addition, this fosters a sense of ownership among those who must implement the plan. The HACCP team might require outside experts who are knowledgeable in the potential microbiological and other public health risks associated with the product and the process. However, a plan which is developed totally by outside sources will likely be erroneous, incomplete, and lacking in support at the local level.

Due to the technical nature of the information required for a hazard analysis, it is recommended that experts who are knowledgeable about the food and process should either participate in or verify the completeness of the hazard analysis and the HACCP plan. These individuals should have the knowledge and experience to correctly (a) identify potential hazards; (b) assign levels of severity and risk; (c) recommend controls, criteria, and procedures for monitoring and verification; (d) recommend appropriate corrective actions when a deviation occurs; (e) recommend research related to the HACCP plan if important information is not known; and (f) predict the success of the HACCP plan.

DESCRIBE THE FOOD AND THE METHOD OF ITS DISTRIBUTION

A separate HACCP plan must be developed for each food product that is being processed in the establishment. The HACCP team must first fully describe the food. This consists of a full description of the food including the recipe or formulation. The method of distribution should be described along with information on whether the food is to be distributed frozen, refrigerated, or shelf stable. Consideration should also be given to the potential for abuse in the distribution channel and by consumers.

IDENTIFY THE INTENDED USE AND CONSUMERS OF THE FOOD

The intended use of the food should be based upon the normal use of the food by end users or consumers. The intended consumers may be the general public or a particular segment of the population, such as infants, the elderly, etc.

DEVELOP A FLOW DIAGRAM WHICH DESCRIBES THE PROCESS

The purpose of the diagram is to provide a clear, simple description of the steps involved in the process. The diagram will be helpful to the HACCP team in its subsequent work. The diagram can also serve as a future guide for others (e.g., regulatory officials and customers) who must understand the process for their verification activities.

The scope of the flow diagram must cover all the steps in the process which are directly under the control of the establishment. In addition, the flow diagram can include steps in the food chain which are before and after the processing that occurs in the establishment. For the sake of simplicity, the flow diagram should consist solely of words, not engineering drawings.

VERIFY FLOW DIAGRAM

The HACCP team should inspect the operation to verify the accuracy and completeness of the flow diagram. The diagram should be modified as necessary.

PRINCIPLE NO. 1:

Conduct a hazard analysis. Prepare a list of steps in the process where significant hazards occur and describe the preventive measures.

The HACCP team next conducts a hazard analysis and identifies the steps in the process where hazards of potential significance can occur. For inclusion in the list, the hazards must be of such a nature that their prevention, elimination or reduction to acceptable levels is essential to the production of a safe food. Hazards which are of a low risk and not likely to occur would not require further consideration. The team must then consider what preventive measures, if any, exist which can be applied for each hazard. Preventive measures are physical, chemical, or other factors that can be used to control an identified health hazard. More than one preventive measure may be required to control a specific hazard. More than one hazard may be controlled by a specified preventive measure.

The hazard analysis and identification of associated preventive measures accomplishes three purposes: First, those hazards of significance and associated preventive measures are identified. Second, the analysis can be used to modify a process or product to further assure or improve safety. Third, the analysis provides a basis for determining CCPs in Principle 2 (Section 4.7).

The hazard analysis procedure differs from the Committee's original HACCP document. This does not negate the validity of current HACCP plans based on the earlier method of hazard analysis. The procedures outlined in this document are recommended for future use. The hazard analysis consists of asking a series of questions which are appropriate to the specific food process and establishment. It is not possible in these recommendations to provide a list of all the questions which may be pertinent to a specific food or process. The hazard analysis should question the effect of a variety of factors upon the safety of the food. Appendix 2-A (this Chapter) lists examples of questions that may be considered during the hazard analysis. The original hazard analysis format is included as Appendix 2-B for comparison.

The hazard analysis must consider factors which may be beyond the immediate control of the processor. For example, product distribution may be beyond the immediate control of the processor, but information on how the food will be distributed could influence, for example, how the food will be processed.

During the hazard analysis, the potential significance of each hazard should be assessed by considering its risk and severity. Risk is an estimate of the likely occurrence of a hazard. The estimate of risk is usually based upon a combination of experience, epidemiological data, and information in the technical literature. Severity is the seriousness of a hazard.

The HACCP team has the initial responsibility to decide which hazards are significant and must be addressed in the HACCP plan. This decision can be debatable. There may be differences of opinion, even among experts, as to the risk of a hazard. The HACCP team must rely upon the opinion of the experts who assist in the development of the HACCP plan.

During the hazard analysis, safety concerns must be differentiated from quality concerns. Hazard is defined as a biological, chemical or physical property that may cause a food to be unsafe for consumption. The term hazard as used in this document is limited to safety. The HACCP team must make the determination whether a potential problem is a safety concern and of its likelihood of occurrence.

Upon completion of the hazard analysis, the significant hazards associated with each step in the flow diagram should be listed along with any preventive measures to control the hazards (step 6 of Fig. 2-1). This tabulation will be used in Principle 2 to determine CCPs.

For example, if a HACCP team were to conduct a hazard analysis for the production of frozen cooked beef patties (Appendix 2-C), enteric pathogens in the raw meat would be identified as a potential hazard. Cooking is a preventive measure which can be used to eliminate this hazard. Thus, cooking would be listed along with the hazard (i.e., enteric pathogens) and the preventive measure as follows:

Step	Identified Hazard	Preventive Measures
5. Cooking	Enteric pathogens	Cooking sufficiently to kill enteric pathogens

PRINCIPLE NO. 2:

Identify the CCPs in the process.

A critical control point is defined as a point, step or procedure at which control can be applied and a food safety hazard can be prevented, eliminated, or reduced to acceptable levels. All significant hazards identified by the HACCP team during the hazard analysis must be addressed.

The information developed during the hazard analysis in Principle No. 1 should enable the HACCP team to identify which steps in the process are CCPs. Identification of each CCP can be facilitated by the use of a CCP decision tree (Fig. 2-2). All hazards which reasonably could be expected to occur should be considered. Application of the CCP decision tree can help determine if a particular step is a CCP for a previously identified hazard.

Figure 2-2. CCP Decision Tree
(Apply at each step of process with an identified hazard.)

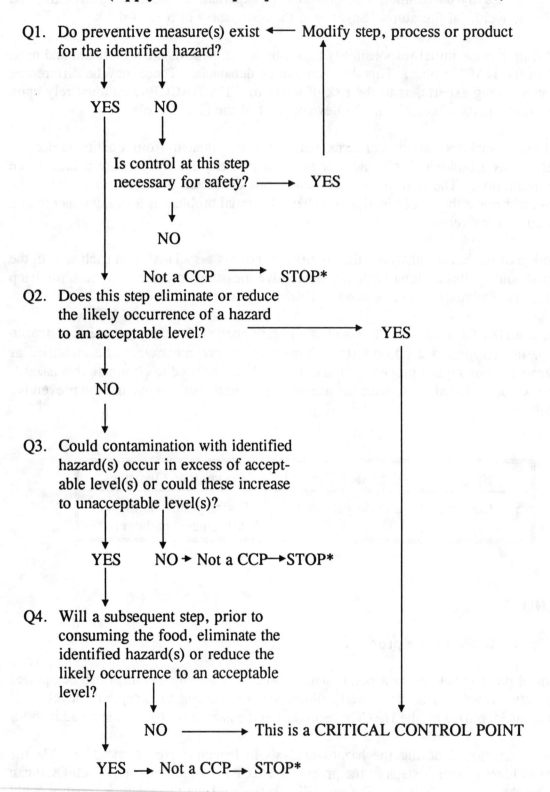

Q1. Do preventive measure(s) exist ◄── Modify step, process or product
for the identified hazard?

YES   NO

Is control at this step
necessary for safety? ───► YES

NO

Not a CCP ───► STOP*

Q2. Does this step eliminate or reduce
the likely occurrence of a hazard
to an acceptable level? ──────────► YES

NO

Q3. Could contamination with identified
hazard(s) occur in excess of acceptable level(s) or could these increase
to unacceptable level(s)?

YES   NO ► Not a CCP──►STOP*

Q4. Will a subsequent step, prior to
consuming the food, eliminate the
identified hazard(s) or reduce the
likely occurrence to an acceptable
level?

NO ────────► This is a CRITICAL CONTROL POINT

YES ──► Not a CCP──► STOP*

---

* Proceed to the next step in the selected process

Critical control points are located at any point where hazards need to be either prevented, eliminated, or reduced to acceptable levels. For example, a specified heat process, at a given time and temperature to destroy a specified microbiological pathogen, is a CCP. Likewise, refrigeration required to prevent hazardous microorganisms from multiplying, or the adjustment of a food to a pH necessary to prevent toxin formation are also CCPs.

Examples of CCPs may include, but are not limited to: cooking, chilling, specific sanitation procedures, product formulation control, prevention of cross contamination, and certain aspects of employee and environmental hygiene.

CCPs must be carefully developed and documented. In addition, they must be used only for purposes of product safety.

Different facilities preparing the same food can differ in the risk of hazards and the points, steps, or procedures which are CCPs. This can be due to differences in each facility such as layout, equipment, selection of ingredients, or the process that is employed. Generic HACCP plans can serve as useful guides; however, it is essential that the unique conditions within each facility be considered during the development of a HACCP plan.

In addition to CCPs, nonfood safety concerns may be addressed at control points. These control points will not be further discussed in this document because they do not relate to food safety and are not included in the HACCP plan.

PRINCIPLE NO. 3:

Establish critical limits for preventive measures associated with each identified CCP.

A critical limit is defined as a criterion that must be met for each preventive measure associated with a CCP. Each CCP will have one or more preventive measures that must be properly controlled to assure prevention, elimination or reduction of hazards to acceptable levels. Each preventive measure has associated with it, critical limits that serve as boundaries of safety for each CCP. Critical limits may be set for preventive measures such as temperature, time, physical dimensions, humidity, moisture level, water activity (A_w), pH, titratable acidity, salt concentration, available chlorine, viscosity, preservatives, or sensory information such as texture, aroma, and visual appearance. Critical limits may be derived from sources such as regulatory standards and guidelines, literature surveys, experimental studies, and experts. The food industry is responsible for engaging competent authorities to validate that the critical limits will control the identified hazard.

For example, an acidified beverage that requires only hot fill and hold as a thermal process may have acid addition as a CCP. If insufficient acid is added, the product would be underprocessed and allow the growth of pathogenic sporeforming bacteria. One preventive measure for this CCP may be pH with a critical limit of pH 4.6. The critical limit for controlling a potential health hazard may be different from criteria associated with quality factors. For example, the product may be of unacceptable quality when the pH exceeds 3.8; however, a health hazard is avoided when the critical limit of pH 4.6 is not exceeded.

In some cases, processing variations may require certain target levels to assure that critical limits are attained. For example, a preventive measure and critical limit may be an internal product temperature of 160°F (71.1°C) during one stage of a process. The oven temperature, however, may be ±5°F (2.8°C) at 160°F; thus an oven target temperature would have to be greater than 165°F (73.9°C) so that no product receives a cook of less than 160°F.

An example for Principle 3 is the cooking of beef patties (Appendix 2-C). The process should be designed to eliminate the most heat-resistant vegetative pathogen which could reasonably be expected to be in the product. Criteria may be required for factors such as temperature, time and meat patty thickness. Technical development of the appropriate critical limits requires accurate information on the probable maximum numbers of these microorganisms in the meat and their heat resistance. The relationship between the CCP and its critical limits for the meat patty example is shown below:

Process Step	CCP	Critical Limits
5. Cooking	YES	Minimum internal temperature of patty: e.g. 145°F Oven temperature: _____ °F Time; rate of heating and cooling (belt speed in rpm):_____ rpm Patty thickness: _____ in. Patty composition: e.g. all beef Oven humidity: _____% RH

PRINCIPLE NO. 4:

Establish CCP monitoring requirements. Establish procedures for using the results of monitoring to adjust the process and maintain control.

Monitoring is a planned sequence of observations or measurements to assess whether a CCP is under control and produce an accurate record (Appendix 2-D) for future use in verification. Monitoring serves three main purposes. First, monitoring is essential to food safety management in that it tracks the system's operation. If monitoring indicates that there is a trend towards loss of control, i.e., exceeding a target level, then action can be taken to bring the process back into control before a deviation occurs. Second, monitoring is used to determine when there is loss of control and a deviation occurs at a CCP, i.e., exceeding the critical limit. Corrective action then must be taken. Third, it provides written documentation for use in verification of the HACCP plan.

An unsafe food may result if a process is not properly controlled and a deviation occurs. Because of the potentially serious consequences of a critical defect, monitoring procedures must be effective. Ideally, monitoring should be at the 100% level. Continuous monitoring is possible with many types of physical and chemical methods. For example, the temperature and time for the scheduled thermal process of low-acid canned foods is recorded continuously on temperature recording charts. If the temperature falls below the scheduled temperature or the time is insufficient, as recorded on the chart, the retort load is retained as a process deviation. Likewise, pH measurement may be performed continually in fluids or by testing of a batch before processing. There are many ways to monitor CCP limits on a continuous or batch basis and record the data on charts. Continuous monitoring is always preferred when feasible. Equipment must be carefully calibrated for accuracy.

Assignment of the responsibility for monitoring is an important consideration for each CCP. Specific assignments will depend on the number of CCPs and preventive measures and the complexity of monitoring. Such individuals are often associated with production (e.g., line supervisors, selected line workers and maintenance personnel) and, as required, quality control personnel. Those individuals monitoring CCPs must be trained in the technique used to monitor each preventive measure; fully understand the purpose and importance of monitoring; have ready access to the monitoring activity; be unbiased in monitoring and reporting; and accurately report the monitoring activity. Personnel assigned the monitoring activity must report the results. Unusual occurrences must be reported immediately so that adjustments can be made in a timely manner to assure that the process remains under control. The person responsible for monitoring must also report a process or product that does not meet critical limits so that immediate corrective action can be taken.

When it is not possible to monitor a critical limit on a continuous basis, it is necessary to establish that the monitoring interval will be reliable enough to indicate that the hazard is under control. Statistically designed data collection or sampling systems lend themselves to this purpose. When using statistical process control, it is important to recognize that critical limits must not be exceeded. For example, when a pH of 4.6 or less is required for product safety, the maximum pH of the product may be set at a target that is below pH 4.6 to compensate for variation.

Most monitoring procedures for CCPs will need to be done rapidly because they relate to on-line processes and there will not be time for lengthy analytical testing. Microbiological testing is seldom effective for monitoring CCPs due to their time-consuming nature. Therefore, physical and chemical measurements are preferred because they may be done rapidly and can indicate the conditions of microbiological control in the process.

Examples of measurements for monitoring include:

- Visual observations
- Temperature
- Time
- pH
- Moisture level

Random checks may be useful for supplementing the monitoring of certain CCPs. They may be used to check incoming pre-certified ingredients, assess equipment and environmental sanitation, airborne contamination, cleaning and sanitizing of gloves and any place where follow-up is needed. Random checks may consist of physical and chemical testing and, as appropriate, microbiological tests.

With certain foods, microbiologically sensitive ingredients, or imports, there may be no alternative to microbiological testing. However, it is important to recognize that a sampling frequency that is adequate for reliable detection of low levels of pathogens is seldom possible because of the large number of samples needed. For this reason, microbiological testing has limitations in a HACCP system, but is valuable as a means of establishing and randomly verifying the effectiveness of control at CCPs (challenge tests, random testing or for troubleshooting).

All records and documents associated with CCP monitoring must be signed or initialed by the person doing the monitoring.

PRINCIPLE NO. 5:

Establish corrective action to be taken when monitoring indicates that there is a deviation from an established critical limit.

The HACCP system for food safety management is designed to identify potential health hazards and to establish strategies to prevent their occurrence. However, ideal circumstances do not always prevail and deviations from established processes may occur. For instances where there is a deviation from established critical limits, corrective action plans must be in place to (a) determine the disposition of non-compliance product, (b) fix or correct the cause of non-compliance to assure that the CCP is under control, and (c) maintain records of the corrective actions that have been taken where there has been a deviation from critical limits. Because of the variations in CCPs for different foods and the diversity of possible deviations, specific corrective action plans must be developed for each CCP. The actions must demonstrate the CCP has been brought under control. Individuals who have a thorough understanding of the process, product and HACCP plan are to be assigned responsibility for taking corrective action. Corrective action procedures must be documented in the HACCP plan.

Should a deviation occur, the plant will place the product on hold pending completion of appropriate corrective actions and analyses. As appropriate, scientific experts and regulatory agencies are to be consulted to determine additional testing and disposition of the product.

Identification of deviant lots and corrective actions taken to assure safety of these lots must be noted in the HACCP record and remain on file for a reasonable period after the expiration date or expected shelf life of the product.

PRINCIPLE NO. 6:

Establish effective recordkeeping procedures that document the HACCP system.

The approved HACCP plan and associated records must be on file at the food establishment. Generally, the records utilized in the total HACCP system will include the following:

1. The HACCP Plan

    - Listing of the HACCP team and assigned responsibilities.
    - Description of the product and its intended use.
    - Flow diagram for the entire manufacturing process indicating CCPs.
    - Hazards associated with each CCP and preventive measures.
    - Critical limits.
    - Monitoring system.
    - Corrective action plans for deviations from critical limits.
    - Recordkeeping procedures.
    - Procedures for verification of HACCP system.

In addition to listing the HACCP team, product description and uses, and providing a flow diagram, other information in the HACCP plan can be tabulated as follows:

Process Step	CCP	Chem. Phys. Biolog. Hazards	Critical Limits	Monitoring Procedures/ Frequency/ Person(s) Responsible	Corrective Action(s)/ Person(s) Responsible	HACCP Records	Verification Procedure/ Person(s) Responsible
1.	Yes or No	1. 2. 3. etc.					

2. Records obtained during the operation of the plan. (See Appendix 2-D.)

PRINCIPLE NO. 7:

Establish procedures for verification that the HACCP system is working correctly.

The National Academy of Sciences (1985)[1] pointed out that the major infusion of science in a HACCP system centers on proper identification of the hazards, critical control points, critical limits, and instituting proper verification procedures. These processes should take place during the development of the HACCP plan. There are four processes involved in verification.

The first is the scientific or technical process to verify that critical limits at CCPs are satisfactory. This process is complex and requires intensive involvement of highly skilled professionals from a variety of disciplines capable of doing focused studies and analyses. The process consists of a review of the critical limits to verify that the limits are adequate to control the hazards that are likely to occur.

The second process of verification ensures that the facility's HACCP plan is functioning effectively. A functioning HACCP system requires little end-product sampling, since appropriate safeguards are built in early in the process. Therefore, rather than relying on end-product sampling, firms must rely on frequent review of their HACCP plan, verification that the HACCP plan is being correctly followed, review of CCP records, and determinations that appropriate risk management decisions and product dispositions are made when process deviations occur.

The third process consists of documented periodic validations, independent of audits or other verification procedures, that must be performed to ensure the accuracy of the HACCP plan. Revalidations are performed by a HACCP team on a regular basis and/or whenever significant product, process or packaging changes require modification of the HACCP plan. The revalidation includes a documented on-site review and verification of all flow diagrams and CCPs in the HACCP plan. The HACCP team modifies the HACCP plan as necessary.

The fourth process of verification deals with the government's regulatory responsibility and actions to ensure that the establishment's HACCP system is functioning satisfactorily.

Examples of verification activities are included as Appendix 2-E.

---

[1] An Evaluation of the Role of Microbiological Criteria for Foods and Food Ingredients. National Academy of Sciences, National Academy Press, Washington, D.C.

APPENDIX 2-A

EXAMPLES OF QUESTIONS TO BE CONSIDERED IN A HAZARD ANALYSIS

The hazard analysis consists of asking a series of questions which are appropriate to each step in a HACCP plan. It is not possible in these recommendations to provide a list of all the questions which may be pertinent to a specific food or process. The hazard analysis should question the effect of a variety of factors upon the safety of the food.

A. Ingredients

1. Does the food contain any sensitive ingredients that may present microbiological hazards (e.g., <u>Salmonella</u>, <u>Staphylococcus aureus</u>); chemical hazards (e.g., aflatoxin, antibiotic or pesticide residues); or physical hazards (stones, glass, metal)?

2. Is potable water used in formulating or in handling the food?

B. Intrinsic Factors

Physical characteristics and composition (e.g., pH, type of acidulants, fermentable carbohydrate, water activity, preservatives) of the food during and after processing.

1. Which intrinsic factors of the food must be controlled in order to assure food safety?

2. Does the food permit survival or multiplication of pathogens and/or toxin formation in the food during processing?

3. Will the food permit survival or multiplication of pathogens and/or toxin formation during subsequent steps in the food chain?

4. Are there other similar products in the market place? What has been the safety record for these products?

C. Procedures Used for Processing

1. Does the process include a controllable processing step that destroys pathogens? Consider both vegetative cells and spores.

2. Is the product subject to recontamination between processing (e.g., cooking, pasteurizing) and packaging?

D. Microbial Content of the Food

1. Is the food commercially sterile (e.g., low acid canned food)?

2. Is it likely that the food will contain viable sporeforming or nonsporeforming pathogens?

3. What is the normal microbial content of the food?

4. Does the microbial population change during the normal time the food is stored prior to consumption?

5. Does the subsequent change in microbial population alter the safety of the food, pro or con?

E. Facility Design

1. Does the layout of the facility provide an adequate separation of raw materials from ready-to-eat foods if this is important to food safety?

2. Is positive air pressure maintained in product packaging areas? Is this essential for product safety?

3. Is the traffic pattern for people and moving equipment a significant source of contamination?

F. Equipment Design

1. Will the equipment provide the time-temperature control that is necessary for safe food?

2. Is the equipment properly sized for the volume of food that will be processed?

3. Can the equipment be sufficiently controlled so that the variation in performance will be within the tolerances required to produce a safe food?

4. Is the equipment reliable or is it prone to frequent breakdowns?

5. Is the equipment designed so that it can be cleaned and sanitized?

6. Is there a chance for product contamination with hazardous substances; e.g., glass?

7. What product safety devices are used to enhance consumer safety?
   - metal detectors
   - magnets
   - sifters
   - filters
   - screens
   - thermometers
   - deboners
   - dud detectors

G. Packaging

1. Does the method of packaging affect the multiplication of microbial pathogens and/or the formation of toxins?

2. Is the package clearly labeled "Keep Refrigerated" if this is required for safety?

3. Does the package include instructions for the safe handling and preparation of the food by the end user?

4. Is the packaging material resistant to damage thereby preventing the entrance of microbial contamination?

5. Are tamper-evident packaging features used?

6. Is each package and case legibly and accurately coded?

7. Does each package contain the proper label?

H. Sanitation

1. Can sanitation impact upon the safety of the food that is being processed?

2. Can the facility and equipment be cleaned and sanitized to permit the safe handling of food?

3. Is it possible to provide sanitary conditions consistently and adequately to assure safe foods?

I. Employee Health, Hygiene and Education

1. Can employee health or personal hygiene practices impact upon the safety of the food being processed?

2. Do the employees understand the process and the factors they must control to assure the preparation of safe foods?

3. Will the employees inform management of a problem which could impact upon safety of the food?

J. Conditions of Storage Between Packaging and the End User

1. What is the likelihood that the food will be improperly stored at the wrong temperature?

2. Would an error in improper storage lead to a microbiologically unsafe food?

K. Intended Use

1. Will the food be heated by the consumer?

2. Will there likely be leftovers?

L. Intended Consumer

1. Is the food intended for the general public?

2. Is the food intended for consumption by a population with increased susceptibility to illness (e.g., infants, the aged, the infirm, immunocompromised individuals)?

APPENDIX 2-B

A. HAZARD ANALYSIS

Rank the food according to hazard characteristics A through F, using a plus (+) to indicate a potential hazard. The number of pluses will determine the risk category. A model diagram outlining this concept is given on page 2-22. As indicated, if the product falls under Hazard Class A, it should automatically be considered Risk Category VI.

Hazard A: A special class that applies to nonsterile products designated and intended for consumption by at-risk populations, e.g., infants, the aged, the infirm, or immuno-compromised individuals.

Hazard B: The product contains "sensitive ingredients" in terms of microbiological hazards.

Hazard C: The process does not contain a controlled processing step that effectively destroys harmful microorganisms.

Hazard D: The product is subject to recontamination after processing before packaging.

Hazard E: There is substantial potential for abusive handling in distribution or in consumer handling that could render the product harmful when consumed.

Hazard F: There is no terminal heat process after packaging or when cooked in the home.

Note: Hazards can also be stated for chemical or physical hazards, particularly if a food is subject to them.

B.   ASSIGNMENT OF RISK CATEGORIES (based on ranking by hazard characteristics):

Category VI: A special category that applies to nonsterile products designated and intended for consumption by at-risk populations, e.g. infants, the aged, the infirm, or immunocompromised individuals. All six hazard characteristics must be considered.

Category V: Food products subject to all five general hazard characteristics. Hazard Characteristics B, C, D, E, and F.

Category IV: Food products subject to four of the general hazard characteristics.

Category III: Food products subject to three of the general hazard characteristics.

Category II: Food products subject to two of the general hazard characteristics.

Category I: Food products subject to one of the general hazard characteristics.

Category O: Hazard Class -- No hazard.

Note: Ingredients are treated in the same manner in respect to how they are received at the plant, <u>before</u> processing. This permits determination of how to reduce risk in the food system.

It is recommended that a chart be utilized that provides assessment of a food by hazard characteristic and risk category. A format for this chart is given as follows:

Food Ingredient or Product	Hazard Characteristics (A, B, C, D, E, F)	Risk Category (VI, V, IV, III, II, I, O)
T	A + (Special Category)*	VI
U	Five +'s (B through F)	V
V	Four +'s (B through F)	IV
W	Three +'s (B through F)	III
X	Two +'s (B through F)	II
Y	One + (B through F)	I
Z	No +'s	O

*Hazard characteristic A automatically is Risk Category VI, but any combination of B through F may also be present.

Appendix 2-B was extracted from Committee's November, 1989 HACCP document.

APPENDIX 2-C

EXAMPLE OF A FLOW DIAGRAM FOR THE PRODUCTION OF FROZEN COOKED BEEF PATTIES

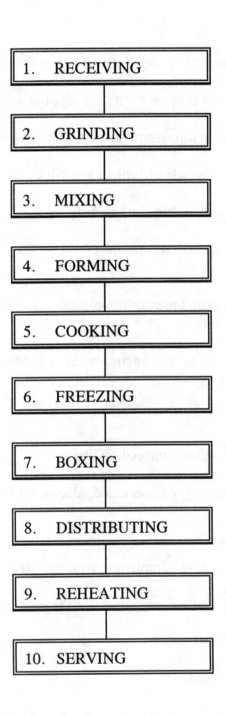

1. RECEIVING
2. GRINDING
3. MIXING
4. FORMING
5. COOKING
6. FREEZING
7. BOXING
8. DISTRIBUTING
9. REHEATING
10. SERVING

APPENDIX 2-D

EXAMPLES OF HACCP RECORDS

A. Ingredients

1. Supplier certification documenting compliance with processor's specifications.

2. Processor audit records verifying supplier compliance.

3. Storage temperature record for temperature sensitive ingredients.

4. Storage time records of limited shelf life ingredients.

B. Records Relating to Product Safety

1. Sufficient data and records to establish the efficacy of barriers in maintaining product safety.

2. Sufficient data and records establishing the safe shelf life of the product, if age of product can affect safety.

3. Documentation of the adequacy of the processing procedures from a knowledgeable process authority.

C. Processing

1. Records from all monitored CCPs.

2. Records verifying the continued adequacy of the processes.

D. Packaging

1. Records indicating compliance with specifications of packaging materials.

2. Records indicating compliance with sealing specifications.

E. Storage and Distribution

1. Temperature records.

2. Records showing no product shipped after shelf life date on temperature sensitive products.

F. Deviation and Corrective Action Records.

G. Validation records and modification to the HACCP plan indicating approved revisions and changes in ingredients, formulations, processing, packaging and distribution control, as needed.

H. Employee training records.

APPENDIX 2-E

EXAMPLES OF VERIFICATION ACTIVITIES

A. Verification procedures may include:

1. Establishment of appropriate verification inspection schedules.

2. Review of the HACCP plan.

3. Review of CCP records.

4. Review of deviations and dispositions.

5. Visual inspections of operations to observe if CCPs are under control.

6. Random sample collection and analysis.

7. Review of critical limits to verify that they are adequate to control hazards.

8. Review of written record of verification inspections which certifies compliance with the HACCP plan or deviations from the plan and the corrective actions taken.

9. Validation of HACCP plan, including on-site review and verification of flow diagrams and CCPs.

10. Review of modifications of the HACCP plan.

B. Verification inspections should be conducted:

1. Routinely, or on an unannounced basis, to assure selected CCPs are under control.

2. When it is determined that intensive coverage of a specific commodity is needed because of new information concerning food safety.

3. When foods produced have been implicated as a vehicle of foodborne disease.

4. When requested on a consultative basis or established criteria have not been met.

5. To verify that changes have been implemented correctly after a HACCP plan has been modified.

C. Verification reports should include information about:

1. Existence of a HACCP plan and the person(s) responsible for administering and updating the HACCP plan.

2. The status of records associated with CCP monitoring.

3. Direct monitoring data of the CCPs while in operation.

4. Certification that monitoring equipment is properly calibrated and in working order.

5. Deviations and corrective actions.

6. Any samples analyzed to verify that CCPs are under control. Analyses may involve physical, chemical, microbiological or organoleptic methods.

7. Modifications to the HACCP plan.

8. Training and knowledge of individuals responsible for monitoring CCPs.

CHAPTER 3

BASIC STEPS IN THE DEVELOPMENT OF HACCP SYSTEMS

by K. E. Stevenson

INTRODUCTION

The description of the HACCP Principles and their application to food manufacturers occupies approximately one dozen pages of the previous chapter. However, the system that is succinctly described in those pages can be highly complex to design and maintain properly. This chapter provides an overview of the development of HACCP systems.

HACCP IS A SYSTEM

HACCP is not a stand-alone, turn-key system which operates by itself. In order to function correctly, HACCP must be installed in an organization in such a manner that a *system* is in place which assures the safety of the foods which are produced under that system. This systematic approach is important because it is needed (a) to evaluate all of the individual components of a particular food production operation, and (b) to ensure that any changes in the system will be evaluated for safety prior to being put into effect.

Thorough documentation is a key element in assuring that the system will operate correctly. This applies to the operating procedures and employee tasks which must be spelled out and documented so that employee-related errors will be minimized, as well as, to the operational records which chronicle the actual operation of the system and any corrective actions which have been conducted.

FOOD SAFETY AND QUALITY

In the past, safety and quality issues have been addressed in various quality control or quality assurance programs and in regulatory compliance programs. One of the strengths of the HACCP system is that it deals only with food safety issues. Thus, the entire focus of HACCP is on food safety, and the safety issues are not diluted out by the various quality, economic and (non-safety) regulatory issues which impact on day-to-day operations in food production facilities.

Due to the "safety only" nature of HACCP, some companies have elected to keep HACCP and QC completely separate. However, in most companies HACCP is associated with the QC function. While HACCP may be placed within a unit which also addresses QC, the actual HACCP Plan and management of HACCP should be kept separate from the actual QC measurements and operations.

ANALYSIS OF COMPONENTS THAT AFFECT SAFETY

The initial stages in developing a HACCP plan involve analyses of various components with respect to food safety. The procedures used for analysis vary according to the various types of components. The following provides comments on evaluation of some of the most important components that affect the safety of food products.

Ingredients and Ingredient Suppliers

Processors may use different strategies in attempts to adequately control hazards associated with raw ingredients. Two of the primary strategies are (a) use of specifications for all raw materials, and (b) use of a mechanism for approving ingredient suppliers.

Food safety-related specifications must be developed for all of the ingredients and packaging materials, as well as, other compounds or chemicals used in the facility. In order to be meaningful, these specifications must be reasonable and must specify achievable limits.

A mechanism must be developed to qualify or approve suppliers. The purpose of qualifying suppliers is to assure that the supplier can meet the specifications and comply with pertinent regulatory requirements. The process of qualifying suppliers usually involves an agreement by the supplier that they can meet the specification, a "letter of guarantee" stating that the supplier guarantees the supplied item meets the specification and applicable regulations, and an on-site inspection of the suppliers facilities and operations evaluating the ability to produce a safe product meeting the specification.

During the past few years, many large companies have required their suppliers to develop HACCP programs in order to become a qualified supplier. Regardless of whether or not a HACCP system is required, companies should conduct periodic tests of incoming supplies in order to verify that the supplied materials conform to specifications.

Processes and Equipment

Many of the most common critical control points (CCPs) are associated with various equipment functions and processes within food plants. Evaluation of plant layout, equipment design, and equipment operation are part of the hazard analysis. Later, careful analyses of processing parameters, including processing targets and variability, are considered in choosing appropriate critical limits (CLs).

Block flow diagrams are commonly used to portray the various pieces of equipment and operations which make up a production line. Based upon a hazard analysis, CCPs are identified on the flow diagram. As an academic exercise, this procedure is often conducted with little accompanying information concerning the equipment and processes. In actual practice, the HACCP Team must have considerable knowledge of, and detailed information concerning, the individual operations and pieces of equipment depicted on the flow diagrams.

Sampling and Testing Procedures

Prior to use in a HACCP Plan, adequate justification should be included for any sampling plan or testing procedure which will be used to obtain data. Any sampling or testing procedures must be presented in detailed, written form. (Note: Document control procedures should be used to assure that everyone is using the current version of each procedure.)

Employee Practices

All facilities should follow a GMP program written and tailored to their needs. Acceptable employee practices must be documented so that employees have specific, written procedures to follow. Personal hygiene, selected access to specific processing areas (traffic control), and strict adherence to other sanitation principles are key elements in the control of operations related to food safety.

In setting up a HACCP Program, special attention should be given to training. An evaluation should be made concerning the training needs at all levels. All employees should be given an overview of HACCP, including information regarding the company's HACCP Policy and Objectives. Specific training of personnel will depend upon their responsibilities and duties related to the HACCP program.

Contract Packers

Contract packers are expected to conform to the same requirements as if they were your own facility. Many large food companies are now requiring that their contract packers operate under a HACCP Plan. This often includes the requirement for submitting a comprehensive HACCP Plan and allowing on-site audits to evaluate the operation of the HACCP Plan. At the present time, this trend is helping spread the use of HACCP to smaller companies throughout the food industry.

Codes, Labels and Instructions

Batch codes, labels and instructions to consumers all contribute to the safety of a product. Codes are used to identify the manufacturing facility, date of production, production line, product, and batch. This information is particularly useful when there are questions concerning product spoilage or recalls of product. Obviously, the smaller the batch size, the less product is in jeopardy if a product withdrawal or recall is necessary. Coding cases with the codes of the individual containers packed therein also will facilitate retrieval of product.

Instructions to the consumer provide important information related to product safety. Appropriate instructions concerning reconstitution, cooking, storage and other handling practices are important in relaying safety-related information to the consumers.

APPLICATION OF HACCP PRINCIPLES

Section III (Chapters 7-11) provides detailed information on tactics for use in developing HACCP Plans and the application of the seven HACCP Principles. Information related to organizing and managing HACCP Programs is provided in Chapter 12.

SECTION II

HAZARDS AND CONTROLS

Biological Hazards and Controls

Chemical Hazards and Controls

Physical Hazards and Controls

CHAPTER 4

BIOLOGICAL HAZARDS AND CONTROLS

by Virginia N. Scott and Lloyd Moberg

FOODBORNE DISEASE IN THE UNITED STATES

INCIDENCE AND IMPACT OF FOODBORNE DISEASE

The Centers for Disease Control estimates that there are 6.5 million cases of foodborne disease with 9100 fatalities annually in the United States (Cohen, 1988 in Archer, 1988). Other estimates are even higher, ranging from 24 to 81 million cases per year (Archer and Kvenberg, 1985). The total yearly cost attributed to foodborne illness in the U. S. is estimated at $1 billion to $10 billion (Todd, 1984).

Foodborne Disease Reported to CDC

The "official" reporting of foodborne disease statistics in the United States began in 1923 with the publication by the Public Health Service of summaries of outbreaks of gastrointestinal illness attributed to milk. In 1938, summaries of outbreaks caused by all foods were added. The Centers for Disease Control (CDC - then the Communicable Disease Center) assumed responsibility for publishing reports on foodborne illness in 1961. Reports on outbreaks of foodborne and waterborne disease come to CDC primarily from state and local health departments. They are also received from federal agencies such as the Food and Drug Administration (FDA), the Department of Agriculture (USDA), the armed forces and occasionally from private physicians.

Foodborne vs. Milkborne and Waterborne Diseases

Early surveillance efforts revealed an association of infant diarrhea and typhoid fever with milk and water, resulting in the requirements for milk pasteurization and chlorination of municipal water supplies (Todd, 1990). This resulted in significant decreases in disease from these sources. In recent years, the numbers of water-related outbreaks have declined (Levine et al., 1990). However, the number of waterborne disease cases in 1987 (Table 4-1) was higher than for any year since CDC and the Environmental Protection Agency began tabulating data in 1971. The number of confirmed outbreaks of foodborne disease remained relatively constant from 1983 to 1987 (Table 4-1). The decrease in confirmed outbreaks of foodborne disease reported in 1987 did not necessarily represent a true decrease. The data reported for foodborne disease outbreaks do not include sporadic cases, which are far more common then cases associated with large outbreaks. Note that the number of sporadic cases of *Salmonella* infection, undoubtedly including some foodborne cases, increased between 1983 and 1987 (Bean et al., 1990).

TABLE 4-1.   OUTBREAKS AND CASES OF FOODBORNE AND WATERBORNE DISEASE[a]

	Foodborne		Waterborne	
Year	Outbreaks[b]	Cases	Outbreaks	Cases
1983	187 (505)	7,904	43	21,036
1984	185 (543)	8,193	27	1,800
1985	220 (495)	22,987	22	1,946
1986	181 (467)	5,804	22	1,569
1987	136 (387)	9,652	15	22,149
1988	- -	-	13	2,128

[a] Adapted from Bean et al., 1990 and Levine et al., 1990
[b] Confirmed (total)

FOODBORNE DISEASE OUTBREAKS

CDC defines a foodborne disease outbreak as an incident in which (1) two or more persons experience a similar illness after ingestion of a common food, and (2) epidemiologic analysis implicates the food as the source of the illness. However, one case of botulism or chemical poisoning constitutes an outbreak (Bean et al., 1990).

Outbreaks are classified by etiologic agent if laboratory evidence of a specific agent is obtained and specified criteria are met. If a food source is implicated epidemiologically but adequate laboratory confirmation of an agent is not obtained, the outbreak is classified as unknown etiology. The etiologic agent was not confirmed in 60% of outbreaks from 1983 to 1987 (Bean et al., 1990), indicating the need for improved investigative techniques to identify known pathogens more frequently and recognize currently unidentified pathogens.

Actual vs. Reported Foodborne Disease Outbreaks

The reported outbreaks of foodborne disease represent only the "tip of the iceberg." The likelihood of an outbreak being recognized and reported to health authorities depends on, among other factors, consumer awareness, physician awareness, disease surveillance activities of state and local health departments, etc. Large outbreaks are more likely to be reported than small ones. Foodborne disease outbreaks associated with food prepared and/or served at restaurants, hospitals and nursing homes are more likely to be recognized than those from family meals at home. Outbreaks involving serious illness, hospitalization or deaths are also more likely to be recognized and reported than those due to pathogens causing mild illness. Foodborne diseases characterized by short incubation periods, such as staphylococcal intoxication, are more likely to be recognized as common source foodborne disease outbreaks than those involving longer incubation periods, such as hepatitis A. Outbreaks involving more common foodborne pathogens are also more likely to be confirmed than those involving less common pathogens, in part because the laboratory may not be knowledgeable in the detection of all foodborne pathogens.

Outbreaks in Selected States

A few states (New York, California, Washington, Hawaii) account for a disproportionate number of outbreaks (Bean et al., 1990). Although this may be due in part to a higher rate of foodborne disease, more likely this is due to better surveillance. Reporting of foodborne disease outbreaks can also be dependent on allocation of state resources. Many health departments have redirected efforts towards the AIDS epidemic, leaving limited resources to address foodborne disease. These occurrences can result in a further underestimation of the size of the foodborne disease problem.

Foods Frequently Involved in Outbreaks

The foods most frequently involved in outbreaks are foods of animal origin. In 48% of the outbreaks between 1973 - 1987, for which the vehicle was known, beef, chicken, eggs, pork, finfish, shellfish, turkey or dairy products were involved (Bean and Griffin, 1990). It is likely that this percentage is low since other outbreaks such as those from ice cream, Mexican food and Chinese food may have been due to food ingredients of animal origin.

Steps Necessary to Cause Foodborne Illness

For a foodborne illness to occur the pathogen or its toxin(s) must be present in the food. In most cases the mere presence of the pathogen is not sufficient for it to cause a foodborne disease; the pathogen must grow to high enough numbers to cause an infection or to produce toxin. Thus, in most instances the food must be capable of supporting growth of the pathogen and the food must remain in the growth temperature range long enough for the organism to multiply and/or produce toxin. Finally enough of the food must be ingested to exceed the threshold of susceptibility of the person ingesting the food (Riemann and Bryan, 1979).

Place Where Illness Acquired

In the 7,219 foodborne disease outbreaks between 1973 and 1987 in which the site of preparation of the implicated food was reported, the sites were commercial or institutional establishments in 79% of the outbreaks and homes in 21% (Bean and Griffin, 1990). In outbreaks occurring between 1983 and 1987, approximately 23% were attributed to food eaten in the home and 44% to food eaten in restaurants, cafeterias or delicatessens (Bean et al., 1990). Food eaten at schools, picnics, churches and camps accounted for another 10-11%. Thus, the data reported to CDC imply foodborne disease is primarily a problem associated with food preparation and storage.

PATHOGENS AND FOODBORNE DISEASE

As noted above, in most cases pathogens must grow (multiply) in foods to appropriate levels to cause foodborne disease. Thus, the food must contain the nutrients required by the organism to grow. The organism must have water, i.e. the available water, or water activity, must be high enough to permit growth. The pH must be in the favorable range and the oxidation-reduction potential must be such that growth can be initiated (Riemann and Bryan, 1979). The food must be free from substances that prevent growth of the pathogen (preservatives, etc.). The food must be at a temperature allowing growth and the organism must be given time to multiply. A number of foodborne pathogens are psychrotrophic, i.e. they are capable of growth at refrigeration temperatures. Table 4-2 shows the minimum temperature for growth of a variety of foodborne pathogens. Most of the organisms capable of growth at refrigeration temperatures grow slowly at low temperatures, requiring extended time to reach high numbers (Table 4-3).

TABLE 4-2. MINIMUM GROWTH TEMPERATURES FOR FOODBORNE PATHOGENS

Organism	Minimum temperature (°C)
Listeria monocytogenes	1
Yersinia entertocolitica	-2
enterotoxigenic *Escherichia coli*	3
Vibrio vulnificus	5
Aeromonas hydrophila	0-5
non-proteolytic *Clostridium botulinum*	3.3
Vibrio parahaemolyticus	5-7
Salmonella	7-10
Bacillus cereus	6-10
Staphylococcus aureus	7-10
proteolytic *C. botulinum*	10
Clostridium perfringens	12

TABLE 4-3. REPRESENTATIVE GENERATION TIMES FOR PSYCHROTROPHIC BACTERIA

Temperature (°C)	Generation time (min.)
25	30
20	75
15	120
10	200
5	1200

Classification of Foods According to pH

Foods can be divided into two major categories: low-acid (pH > 4.6) and acid (pH ≤ 4.6). These categories were established based upon the growth of *Clostridium botulinum*. The minimum pH for growth of *C. botulinum* in foods is generally accepted as 4.8, although it has been shown to grow as low as pH 4.0 in strictly controlled laboratory environments (Raatjes and Smelt, 1979; Smelt et al., 1982; Tanaka, 1982; Young-Perkins and Merson, 1986). The minimum pH for growth of *Salmonella* in laboratory media is 4.0 (Chung and Goepfert, 1970). The minimum pH for growth of *Listeria monocytogenes* in laboratory media is between 4.5 and 5.0, depending on the acidulant (Conner et al., 1990; Sorrells et al., 1989).

Types of Foodborne Disease

Foodborne disease can be classified as either infections or intoxications. Infections are caused by viable pathogenic microorganisms entering the body and colonizing, and the body's reacting to the organism or toxins produced by it (Riemann and Bryan, 1979). There are two types of foodborne infections (Riemann and Bryan, 1979). One type results from penetration of the intestinal mucosa by the infecting organism and its subsequent multiplication therein (*Salmonella, Shigella*), or its multiplication in other tissues (hepatitis A, *Trichinella spiralis*). A second type results from release of enterotoxins by an infecting organism as it multiplies, lyses, or sporulates in the intestinal tract (*Clostridium perfringens, Vibrio cholerae*).

An intoxication is caused by the ingestion of toxins. These may be found naturally in certain plants and animals (e.g. poisonous mushrooms), or may be metabolic products of certain bacteria (botulinum toxin, staph enterotoxin), molds (mycotoxins) or algae/dinoflagellates (saxitoxin).

The CDC classifies foodborne disease outbreaks as bacterial, chemical, parasitic or viral. The majority of cases of foodborne disease are caused by bacterial agents (Table 4-4). In the United States, the most common bacterial agents of foodborne disease include *Salmonella, S. aureus, C. perfringens* and, to a lesser degree, *Bacillus cereus, Shigella, Campylobacter* and *C. botulinum*. In foodborne disease outbreaks in which the etiologic agent was identified, these organisms accounted for 93% of the outbreaks and 94% of the cases between 1973 and 1987 (Bean and Griffin, 1990). *Campylobacter, Escherichia coli* O157:H7 and *Listeria monocytogenes* have only recently been recognized as foodborne disease organisms. However, since *Campylobacter* outbreaks were first reported to the CDC system in 1980, *Campylobacter* has accounted for up to 6% of cases and outbreaks annually (Bean and Griffin, 1990).

Hepatitis A and Norwalk viruses account for most foodborne illness from viruses. Although viruses have accounted for less than 10% of foodborne disease outbreaks and cases, they are probably a much more important cause of foodborne disease than the data suggest. As diagnostic capabilities for viruses increase, the proportion of foodborne disease attributed to them will also increase.

The most common foodborne parasites are *T. spiralis* and *Giardia lamblia*. The proportion of foodborne disease from parasites has decreased from 34% in 1974-75 to 15% in 1985-87. Histamine (produced by bacterial decomposition of scombroid fish) and ciguatoxin accounted for about half the cases of chemical foodborne disease of known etiology between 1983-1987 (Bean and Griffin, 1990). Most other cases were due to heavy metals, mushroom poisoning and paralytic shellfish poisoning.

TABLE 4-4. ETIOLOGIC AGENTS OF FOODBORNE DISEASE 1973-1987[a]

Agent	Outbreaks (%)	Cases (%)
Bacteria	66	87
Viruses	5	9
Parasites	5	1
Chemicals	25	4

[a] Bean and Griffin (1990)

FACTORS CONTRIBUTING TO FOODBORNE DISEASE OUTBREAKS

Improper Storage/Holding Temperature

Improper storage or holding temperature is the most common factor contributing to bacterial foodborne illness. Foodborne disease organisms will grow in foods held at temperatures between 5°C and 55°C; most bacterial pathogens grow very rapidly at temperatures between 25 and 40°C. Thus, hot foods that are not rapidly cooled for storage or held hot enough prior to consumption may be at temperatures in the "danger zone" (allowing bacterial growth) for sufficient time to produce enough organisms or toxin to cause illness. Foods prepared several hours ahead of time and in large quantities are sometimes improperly cooled (e.g. refrigerated in large, deep containers) or held at improper temperatures (e.g. on steam-tables or in ovens), resulting in outbreaks of foodborne disease. Improper holding temperature is a frequent contributing factor in outbreaks attributed to *C. perfringens, B. cereus, S. aureus* and *Salmonella*.

Inadequate Cooking

Inadequate cooking represents a hazard since cooking is relied upon to destroy many foodborne disease organisms and toxins. Undercooking poultry can lead to illness from *Salmonella* or *Campylobacter*, and improperly processing canned food can result in botulism (the latter is primarily a problem with home-canned or home-prepared foods). Similarly, undercooked seafood can result in illness from *Vibrio parahaemolyticus* and *V. cholerae*, and undercooked pork or bear meat can result in trichinosis.

Poor Personal Hygiene

Many foodborne disease organisms are transferred by the fecal-oral route. Infected food handlers with poor personal hygiene transfer organisms to the food. This is a major contributing factor in outbreaks due to viruses (hepatitis A, Norwalk) and bacteria such as *Shigella*. *S. aureus* may be transferred from the skin or nares of food handlers and, if given sufficient time to grow, may produce toxin in the food.

Cross-contamination

Foodborne pathogens can be transferred from raw product to utensils and equipment, which, if then used for cooked or other ready-to-eat foods, can transfer the pathogens and lead to illness. Cutting boards, slicers, mixers and grinders with hard-to-clean areas are particular problems. Utensils and equipment used in the preparation of raw products should never be used for cooked products without thorough cleaning and sanitizing. Cross-contamination can also occur when cooked foods are stored with raw product, particularly raw foods of animal origin.

Improper Reheating

If pathogens survived the cooking process or cross contamination occurs followed by temperature abuse, the number of organism present may survive reheating the food. This is particularly of concern when leftovers are warmed rather than thoroughly reheated.

Poor Storage Practices

If cooked product is stored with raw product or ingredients, contamination may occur. For example, when frozen raw meats are thawed in a refrigerator containing other foods, drip from the thawing meat can potentially contaminate cooked or ready-to-eat foods; this may cause illness when consumed. Also, storing chemicals and foods together can result in contamination of foods that may cause illness.

PREVENTION OF FOODBORNE DISEASE

There are three major ways of preventing foodborne disease: prevent contamination of foods; destroy foodborne disease agents that may be present in foods; and/or prevent foodborne disease agents from growing in foods

Prevent Contamination of Foods

It should be assumed that raw foods may contain pathogens (e.g. *Salmonella* on poultry, *C. botulinum* on vegetables). However, it is possible to prevent contamination of many of the foods we eat. The use of good personal hygiene practices in food preparation will help prevent foodborne disease from organisms such as viruses, *Salmonella*, *Shigella* and *S. aureus*.

Raw foods should be handled separately from cooked and ready-to-eat foods to avoid cross contamination. Utensils, equipment and work surfaces used for raw foods should be thoroughly cleaned and sanitized prior to using for cooked or ready-to-eat foods. Equipment and utensils should be clean and made from appropriate materials to avoid contamination from toxic materials.

Destruction of Foodborne Disease Agents

Many foodborne disease organisms will be destroyed by proper cooking. In general, raw animal products such as eggs, fish, lamb and beef should be cooked to 63°C or above to destroy pathogens (FDA, 1988). Freezing can be used to destroy parasites in fish and meat, but it has little effect on bacterial pathogens in food. While irradiation can be used in some cases to destroy pathogens (e.g. on raw poultry), this procedure is rarely used because of consumer concerns. Acids and preservatives sometimes kill certain microorganisms, however, in most cases they are used to prevent growth rather than to kill.

Prevent Multiplication of Foodborne Disease Agents

As indicated previously, the presence of certain foodborne disease agents at any level is a hazard. However, many other organisms must multiply to large numbers to cause disease. For example, *S. aureus* must reach levels of 10^5 - 10^6 to produce enough toxin to cause disease. *C. perfringens* and *V. cholerae* must also be present in high numbers (approximately 10^6/g) to cause illness. Thus, storing and preparing foods under conditions that prevent growth (multiplication) is a primary means of preventing foodborne disease.

Although *Y. enterocolitica* can grow at temperatures as low as -2°C, freezing generally prevents growth of all foodborne disease organisms. Proper refrigeration temperatures ($\leq$4°C) will prevent multiplication of most foodborne disease organisms (Table 4-2) and slow the multiplication of others (Table 4-3); the lower the temperature, the slower the growth rate.

It is important to lower the temperature of foods rapidly to keep microorganisms from growing. A good rule of thumb is to reduce the temperature from 60°C to 4°C (140°F to 40°F; the "danger zone" for microbial growth) in $\leq$4 hours. In order to cool foods rapidly, they should be placed in shallow pans or divided into small portions. If cooled in a refrigerator, food should be loosely covered and arranged to allow air to flow around the containers. Foods can also be rapidly chilled by stirring the food in a container in an ice water bath or by using special equipment for rapid chilling.

Decreasing the pH and/or the water activity of a food or judicious use of preservatives can prevent or slow the growth of foodborne pathogens. Combining subinhibitory levels of several factors can be used effectively to control pathogens, particularly under refrigeration conditions.

Holding foods at elevated temperatures ($>$60°C; 140°F) can also prevent growth of foodborne pathogens. Care should be taken to see that all parts of the food are above 60°C. For example, on steam trays where the heat source is beneath the trays, the temperature at the surface should be kept at 60°C or above.

MISCELLANEOUS FOODBORNE DISEASE

Giardia lamblia

Although *Giardia* is primarily a waterborne disease organism, it can sometimes be foodborne. In 1983-1987 there were 3 foodborne outbreaks of giardiasis involving 41 cases (Bean et al., 1990). In contrast, in 1986-1988 there were 9 waterborne outbreaks of giardiasis involving 1,169 cases (Levine et al., 1990). *Giardia* may become foodborne when contaminated water is used to wash foods, such as fruits and vegetables. In addition, poor personal hygiene by an infected food handler can result in foodborne transmission of *Giardia*.

Foodborne Viruses

Since viruses are obligate intracellular parasites they cannot multiply in food. Foodborne viral disease results from fecal contamination of food, generally due to poor personal hygiene of a food handler (IFT, 1988).

Hepatitis A and Norwalk agent are the viral agents most commonly transmitted through foods. From 1983 to 1987, hepatitis A caused 71% of the outbreaks due to viruses (Bean et al., 1990). The low number of reported outbreaks due to Norwalk agent and other viruses reflects the limitations of current laboratory techniques for detecting viruses and viral infections (Bean et al., 1990).

CONCERNS ABOUT AIDS

The Human Immunodeficiency Virus (HIV) which causes the disease AIDS (Acquired Immune Deficiency Syndrome) is a severe public health problem. AIDS has never been shown to be transmitted by food or drink (Khan, 1988; IFT, 1988). Individuals who are known to be infected with the virus can handle food safely if they observe basic sanitation precautions for food handling and take care to avoid injury when preparing food. As with any food handler, should an injury occur, food contaminated with blood should be discarded for aesthetic as well as safety reasons. Employees should be restricted from handling food if they have evidence of infection or illness that would otherwise require that they not handle food (CDC, 1985).

PATHOGENS & INDICATORS: CHARACTERISTICS & CONTROLS

The remaining sections of this chapter will discuss the microbiological hazards that are responsible for foodborne illness and what can be done to ensure their control. To establish a comprehensive HACCP program, microbiological hazards need to be carefully identified and evaluated. An overview of microbiological hazards includes the following:

- Classification with respect to health hazards;
- Examples of significant microbiological hazards and their characteristics; and
- Methods to be used to control microbiological hazards.

This material on microbiological hazards provides only a fraction of the information needed to develop an acceptable HACCP program. A microbiologist who is knowledgeable about the particular industry and manufacturing process should be part of the HACCP team which establishes the HACCP program.

CLASSIFICATION OF MICROORGANISMS ACCORDING TO DEGREE OF HEALTH HAZARD

Not all microorganisms are created equal when one measures the potential for causing foodborne illness. The potential for causing illness, or the type of hazard a microbe presents, ranges from severe to none, with every variation between these extremes. The International Commission on Microbiological Specifications for Foods (ICMSF) has arbitrarily divided such microorganisms into various categories to help in developing a risk assessment model (ICMSF, 1986). In the simplest sense, microorganisms can present one of two types of hazards with regard to health: no direct health hazard or a health hazard.

Microbes which do not present a direct health hazard can still be important in the general contamination of a product causing reduced shelf-life and spoilage. Microorganisms which represent health hazards have been arbitrarily divided into four groups (ICMSF, 1986; NAS/NRC, 1985):

- Severe, direct health hazards
- Moderate, direct hazards with potentially extensive spread
- Moderate, direct hazards with limited spread
- Low, indirect hazards, or microbes that serve as indicators of a potential for a more severe hazard or condition.

Table 4-5 provides the classifications for some of the most common foodborne disease agents.

As with most situations, there is usually more than one factor that may influence a particular risk. The type of hazard that a foodborne microorganism may present is further influenced by handling conditions to which the food is subjected. Food that is handled in a manner that destroys the microorganisms (i.e. cooking) reduces the risk or degree of the hazard of foodborne illness. Food that is maintained in a steady state (i.e. frozen) where the microbes cannot grow, causes no change in the risk or degree of hazard. Food that is handled such that the microorganisms are allowed to proliferate (i.e. improper thawing) may increase the hazard.

When specific situations are evaluated with these two parameters (the type of microbiological hazard and the conditions of food handling), different levels of risk become apparent. Table 4-6 displays the various levels of risk (called "cases") that are possible. Cases in which the more serious microbial hazards are allowed to increase (Cases 12 and 15) obviously represent a higher degree of risk than those in which less serious microbial hazards are reduced (Cases 4 and 7). Generally, with a higher degree of risk, more care is required in the processing and handling of the food item to prevent a microbiological hazard from developing into a health hazard. With HACCP, food products with high degrees of risk would have more stringent monitoring conditions than food products with little or no health hazard risk. The 15 cases shown in Table 4-6 have different levels of sampling associated with them. Those cases with the highest risk would require a higher level of sampling than low risk cases. A more detailed discussion of sampling for various microbiological risks can be found in the Chapter on Critical Limits and Monitoring.

TABLE 4-5. CLASSIFICATION OF FOODBORNE DISEASE AGENTS[a]

■ **Severe, Direct Health Hazards:**

- *Clostridium botulinum*
- *Shigella dysenteriae*
- *Listeria monocytogenes*
- *Escherichia coli O157:H7*
- *Salmonella typhi*
- *Salmonella paratyphi A & B*
- *Brucella abortus*; *Brucella suis*
- *Mycobacterium bovis*
- *Vibrio vulnificus*
- Hepatitis A virus
- Fish and shellfish toxins
- Certain Mycotoxins (aflatoxin)

■ **Moderate Hazards with Potentially Extensive Spread:**

- *Salmonella* spp.
- pathogenic *Escherichia coli* (e.g., enterotoxigenic)
- *Streptococcus pyogenes*
- *Shigella* spp.

■ **Moderate Hazards with Limited Spread:**

- *Staphylococcus aureus*
- *Clostridium perfringens*
- *Bacillus cereus*
- *Vibrio parahaemolyticus*
- *Coxiella burnetii*
- *Yersinia enterocolitica*
- *Campylobacter fetus*
- *Trichinella spiralis*
- histamine (from microbial decomposition of scombroid fish)

[a]Adapted from ICMSF, 1986.

TABLE 4-6. PLAN STRINGENCY (CASE) IN RELATION TO DEGREE OF HEALTH HAZARD AND CONDITIONS OF USE[a]

Type of Hazard	Conditions in which food is expected to be handled		
	Reduce Degree of Hazard	Cause No Change in Hazard	May Increase Hazard
No Direct Health Hazard (Spoilage)	Case 1	Case 2	Case 3
Low, Indirect (Indicators)	Case 4	Case 5	Case 6
Moderate, Direct, Limited Spread	Case 7	Case 8	Case 9
Moderate, Direct, Potentially Extensive Spread	Case 10	Case 11	Case 12
Severe, Direct	Case 13	Case 14	Case 15

[a]Adapted from ICMSF, 1986.

INFECTIOUS DOSE

The numbers of microorganisms needed to cause illness varies with the specific strain of the microorganism and the susceptibility of the host. A child may be more susceptible and therefore fewer numbers will be required to cause illness than for an adult. Likewise, hosts that are elderly, debilitated, suffering from other illnesses or injuries, immunocompromised, or somehow less resistant, may also become ill when exposed to fewer pathogenic microorganisms than would be required to cause illness in a healthy adult. With toxigenic microorganisms, the level of toxin needed to cause the disease is more important than the level of the microorganism.

Table 4-7 provides examples of the varying doses of microorganisms needed to cause disease. The challenge doses needed to evoke a clinical response of disease in adult humans is shown with a variety of pathogenic microorganisms. With the exception of *Shigella dysenteriae*, the infective dose appears rather high. However, remember that contamination of a food with a very low level of these microbes can still cause illness, if the food is subsequently mishandled. In a HACCP program, identification of microbiological hazards is equally as important as the identification of proper critical control points, controls and monitoring procedures to assure that the controls are functioning.

Table 4-7. CLINICAL RESPONSE OF ADULTS TO VARYING CHALLENGE DOSES OF ENTERIC PATHOGENS[a]

Organism	Challenge Dose
	(cells)
Shigella dysenteriae	10^1 - 10^4
Shigella flexneri	10^2 - 10^9
Vibrio cholerae	10^3 - 10^9
Salmonella typhi	10^4 - 10^9
Salmonella species (non-typhi)	10^5 - 10^{10}
Escherichia coli (pathogenic types)	10^6 - 10^{10}
Clostridium perfringens	10^8 - 10^9
Yersinia enterocolitica	10^9

[a]Adapted from Riemann and Bryan, 1979.

CHARACTERISTICS OF COMMON FOODBORNE PATHOGENS

Presentation of detailed information concerning the characteristics, properties and diseases caused by all of the foodborne pathogens is beyond the scope of this manual. Furthermore, due to the similar nature of many of the non-sporeforming bacterial foodborne pathogens, control of the more common types of foodborne disease agents will also control vegetative cells of other bacterial pathogens.

Tables 4-8 through 4-13 present information on six bacteria which are the most common causes of bacterial foodborne disease, or of the most concern, from the standpoint of control in the food processing industry. Each of these microorganisms will be discussed from the standpoint of the following:

- The disease caused by the microorganism or its toxin(s)
- The source (reservoir) of the microorganism
- The most common method(s) of transmission
- The characteristics important in control

Additional information on hazardous microorganisms can be found in cited references (ICMSF, 1986; NAS/NRC, 1985; Riemann and Bryan, 1979).

Table 4-8. *Clostridium botulinum*

Disease, Symptoms and Onset:

Botulism (from the botulinum toxin produced by the organism). A severe intoxication resulting from the ingestion of preformed toxin. Blurred or double vision, ptosis, dry mouth, difficulty swallowing, paralysis of respiratory muscles. Vomiting and diarrhea may be initially present. Symptoms develop 12-36 hrs after eating contaminated food (sometimes days). Unless adequately treated (antitoxin, respiratory support), fatality rate high. Recovery may be slow (months, rarely years).

Source:

Soil, marine sediment, and the intestinal tract of animals, including fish. Vegetables and grains will contain *C. botulinum* spores.

Transmission:

Toxin must be ingested to cause disease. Spores must germinate and grow to vegetative cells to produce toxin. Spores are ubiquitous. Spores must be assumed to be present on all foods, including frozen and refrigerated foods.

Characteristics of Microorganism:

Sporeforming, gram + rods. Spores are extremely heat resistant. Controlled retort processing is necessary to destroy. Toxin is destroyed by heat (boiling for 5 min.).

Organisms grows best under anaerobic or reduced oxygen conditions. Non-proteolytic types can grow at low temperatures (to 3°C; 37°F); most proteolytic types can grow to 10°C (50°F).

Low pH prevents growth of microbe. No growth below pH 4.6 demonstrated in commercial food.

Spores can germinate and grow in most low-acid foods under anaerobic conditions. Primarily associated with underprocessed home-prepared foods. Note: Also can cause problems if competitive microbes are destroyed and then the product is subjected to temperature abuse, e.g. frozen pot pies, baked potatoes.

Best control: Retort product to destroy spores, add inhibitor to spore germination, low pH, low a_w (water activity), temperature control, proper heating of food.

Table 4-9. *Listeria monocytogenes*

Disease, Symptoms and Onset:

Listeriosis. An acute meningoencephalitis with or without associated septicemia. Characterized by sudden fever, intense headache, nausea, vomiting, delirium and coma (in elderly, immunocompromised, infants, and pregnant women). May cause abortion. Fatality rate about 30%. In normal host, may cause few symptoms to an acute, mild, febrile illness with flu-like symptoms. Incubation period is unknown; likely 3 days to several weeks.

Source:

Infected domestic and wild animals, fowl and humans. Found in water and mud. Seasonal use of silage is followed by an increase in listeriosis in animals. Assume milk, agricultural commodities contaminated.

Transmission:

Associated with the consumption of contaminated vegetables and dairy products. In neonates, transmission from mother to fetus *in utero*.

Characteristics of Microorganism:

Nonsporeforming, gram + rod, killed by pasteurization temperatures (71.7°C for 15 sec).

Grows under aerobic and anaerobic conditions. Able to grow at refrigeration temperatures (to 1°C; 34°F).

Low pH (< 4.6) prevents growth of organism.

Extremely hardy in comparison to most vegetative cells. Withstands repeated freezing and thawing. Survives for prolonged periods in dry conditions.

Live organisms must be ingested to cause disease. Infectious dose related to susceptibility of host. Susceptible hosts may succumb to as few as 100 cells; healthy, non-susceptible hosts may withstand 10,000,000 cells.

Best control: Low pH, proper heat treatment, avoidance of recontamination, proper temperature control, low a_w.

Table 4-10. *Salmonella* spp.

Disease, Symptoms, and Onset:

Salmonellosis. An acute gastroenteritis characterized by sudden onset of headache, abdominal pain, diarrhea, nausea and vomiting. Fever is always present. Dehydration may be severe. In some instances may cause death. Incubation period is 6 to 72 hours, usually about 12 to 36 hours.

Source:

Intestinal tract of domestic and wild animals, and humans.

Transmission:

Ingestion of the organism in food from infected animals or contaminated by the feces of an infected animal or person. Includes raw eggs, raw milk, meat and poultry. Infectious dose may be only a few cells (100 to 1000), but generally is much higher.

Characteristics of Microorganism:

Nonsporeforming, gram - rod, killed by mild heat (above 60°C).

Grows under aerobic and anaerobic conditions. Grows in a temperature range of 5.2 to 47°C. Optimum temperature for growth is 35 to 37°C.

Low pH ($<$ 4.6) prevents growth; optimum pH for growth is 6.5 to 7.5.

Survives well in frozen or dry state. Organisms in dry state (and in foods with relatively low water activities) are more resistant to heat.

Over 2000 serovars of salmonellae are known.

Best Control: Thorough cooking of food, avoid recontamination, low pH, proper hygiene of food handlers

Table 4-11. *Escherichia coli* - Pathogenic Types, Including O157:H7

Disease, Symptoms and Onset:

Gastroenteritis. Diarrhea, may be bloody, and fever. Strains that cause diarrhea may be invasive, enteropathogenic, or enterotoxigenic. Incubation period is generally 12 to 72 hr from ingestion of food.

E. coli O157:H7 causes hemorrhagic colitis; severe abdominal cramps followed by bloody diarrhea, vomiting, nausea and low-grade fever. May develop into hemolytic uremic syndrome (HUS) in children (kidney failure); symptoms appear 3-4 days after ingestion of food.

Source:

Intestinal tract of humans and animals. Infected persons are often asymptomatic.

Transmission:

Major mode of transmission is fecal contamination of food or water. Cross contamination. Person-to-person spread has been demonstrated. Poor handwashing in day care and nursery after patient contact has contributed to spread of disease. Infectious dose is unknown, but carriers shed large numbers of microorganisms. Transmission of *E. coli* O157:H7 primarily associated with undercooked ground beef.

Characteristics of Microorganism:

Nonsporeforming, gram - rod, killed by mild heat (above 60°C).

Grows under aerobic or anaerobic conditions. Grows well in moist, low-acid foods at temperatures ≥ 7°C. Optimum temperature for growth 35-37°C.

Low pH (< 4.6) will prevent growth.

Difficult to differentiate from non-pathogenic *E. coli* in usual microbiological testing.

Control: Proper cooking and reheating of foods, proper refrigeration (≤ 4°C), good sanitation and personal hygiene, low pH, low a_w.

Table 4-12. *Staphylococcus aureus*

Disease, Symptoms, and Onset:

Staphylococcal food poisoning. An intoxication of abrupt onset characterized by severe nausea, cramps and vomiting. Often accompanied by diarrhea. Deaths rare. Duration of illness one to two days. Onset of symptoms between 1 and 6 hours after consumption of food; generally, 2 to 4 hours.

Source:

Usually humans, organism harbored in nasal passages and on skin. Occasionally from cows with infected udders.

Transmission:

Ingestion of food containing staphylococcal enterotoxin. Microbe multiplies in food and produces a heat-stable enterotoxin. Food handlers commonly contaminate foods that do not undergo adequate heating or refrigeration to prevent growth of microbe (e.g., sandwiches, custards, salad dressings, pastries, sliced meats). As little as 2 hours at unrefrigerated temperatures may allow sufficient growth and toxin production.

Characteristics of Microorganism:

Nonsporeforming, gram + cocci, killed by mild heat (above 60°C). Enterotoxins are very heat stable, and will withstand boiling for prolonged periods. May not be inactivated during normal retorting process.

Grows in either aerobic or anaerobic conditions. Temperature growth range is 6.7 to 45.4°C. Optimum growth temperature is 37 to 40°C.

pH growth range 4.5 to 9.3. Enterotoxin generally not produced below pH 5.2.

Organisms resistant to high salt (up to 15%).

Large numbers of cells (>500,000) per gram food needed in order to produce sufficient amounts of enterotoxin to result in illness.

Control: Proper hygiene, proper refrigeration of foods (<4°C) or proper holding when hot (>60°C) of perishable foods, exclusion of food handlers with boils, sores, abscesses.

Table 4-13. *Clostridium perfringens*

Disease, Symptoms and Onset:

Perfringens food poisoning. A gastroenteritis characterized by diarrhea and nausea. Vomiting and fever are usually absent. Mild disease of short duration (one day). Incubation period from 6 to 24 hours, usually from 10 to 12 hr.

Source:

Soil. Intestinal tract of healthy persons and animals (cattle, pigs, poultry, fish).

Transmission:

Ingestion of food, contaminated by soil or feces, which had been held under conditions permitting growth of organisms. Usually inadequately heated or reheated meats, stews or gravies. Spores survive normal cooking temperatures and germinate and grow during mishandling after cooking. Dose causing disease generally >500,000 cells per gram. Enterotoxin is produced in the gut resulting in symptoms. Common in cafeterias, food service establishments that have inadequate facilities for cooking and refrigeration of large amounts of food, or inadequate hot holding of food.

Characteristics of Microorganism:

Sporeforming, gram + rod. Spores survive normal cooking procedures, including boiling.

Grows well anaerobically and in reduced oxygen conditions. Temperature range for growth is 12° to 50°C. Optimum growth temperature is 43° to 45°C.

Slow cooling and non-refrigerated storage of cooked meat and poultry permit growth to high numbers needed for infection. Can grow in foods placed on steam tables if food is not adequately heated (≥60°C). Inadequate reheating allows organisms to survive.

Control: Proper heating, reheating, and cooling of cooked, perishable foods. Large quantities of food should be distributed in shallow pans for proper cooling during refrigeration.

INDICATOR ORGANISMS (AND RELATED TESTS)

Indicator organisms <u>do not represent a direct health hazard</u> in a food product. However, they do serve to indicate that the potential is present for a health hazard to exist. Generally, these organisms or related tests may signal:

- The possible presence of a pathogen or toxin; or
- The possibility that faulty practices occurred during production, processing, storage and distribution (NAS/NRC, 1986).

Indicator organisms have been used to indicate fecal contamination or lack of cleanliness in food preparation. The coliform bacteria and *E. coli* are two indicators commonly used for this purpose. For example, pasteurized milk should not contain these organisms; their presence would indicate either inadequate processing or recontamination after processing. Since pathogenic organisms would come from the same source as the indicators (e.g., fecal material would also potentially contain *Salmonella* spp.), detection of coliforms or *E. coli* would indicate a possible health hazard. Generally, the detection of indicator organisms is easier than the associated pathogens (i.e. indicators are in greater numbers, detection methodology is faster).

Common indicator organisms or related tests include the following:

- Aerobic Plate Count (APC)
- Variations of APC
    - Thermoduric Count
    - Psychrotrophic Count
    - Thermophilic Count
    - Proteolytic Count
    - Lipolytic Count
- Staphylococci
- Coliforms
- Fecal Coliforms
- *Escherichia coli* (non-pathogenic types)

Occasionally, direct counts and chemical tests can also be used as indicators of the acceptability of ingredients or foods. Examples such counts and tests include the following:

- Direct microscopic count (e.g. raw milk)
- Howard mold count (e.g. tomato paste)
- Dye reduction test (e.g. milk)
- pH (e.g. indicator of fermentation activity)
- Indol (e.g. seafoods)
- Diacetyl (e.g. citrus and vegetable conc.)
- Histamine (e.g. scombroid fish)
- Phosphatase test (e.g. milk pasteurization)

Remember, indicator organisms and related tests serve to indicate the suitability of a food or ingredient for a desired purpose.

SPOILAGE MICROORGANISMS

Spoilage microorganisms do not represent a health hazard. They are associated with spoilage and the economic loss of an ingredient or food. Spoilage organisms affect the quality of the food product, not the safety of a food.

Spoilage organisms represent a broad group of microbes. Generally, they are specific to the type of food and the technology of processing. Examples of food groups and their related spoilage microorganisms include the following:

- Refrigerated foods - psychrotrophs
- Juice concentrates - osmophilic yeasts
- Fermented foods - acid tolerant lactic acid bacteria and yeast
- Meat products - psychrotrophic pseudomonads
- Hot-filled juices - heat resistant molds

Microorganisms that spoil foods can do so at various times in the process: Before and during preparation or processing; under normal conditions of intended use; under unusual circumstances, i.e. if present, and not destroyed or controlled by normal processing techniques.

The normal flora of foods would contain spoilage microorganisms. Their control would depend upon proper handling of the food or ingredient. However, spoilage flora may eventually grow to levels that will spoil food even with the best control systems (e.g., note the limited shelf life of refrigerated foods).

Spoilage microorganisms do not represent a hazard and thus are not of immediate concern when setting up a HACCP program. While they may cause economic loss of a food product, they are not a threat to the health of a consumer.

CONTROL OF MICROORGANISMS

To determine the best means to control microorganisms and microbial toxins, one must first understand the characteristics of the microorganisms that may be amenable to some form of control (Tompkin and Keuper, 1982). Thus, determine what allows them to be present or survive in a food. Various factors that influence the presence and/or level of microbes include

- Source of microbe: naturally-occurring on ingredient, in-process contaminant from equipment, food handlers, etc;
- Temperature of growth: optimum and range;
- Heat resistance: vegetative cells, spores, toxins;
- Sensitivity to acidity: pH limits for growth, optimum and range;
- Sensitivity to low moisture: a_w limit for growth, optimum and minimum;
- Sensitivity to preservatives;
- Influence of oxygen: aerobic, anaerobic, facultative, microaerophilic;
- Sensitivity to unique conditions: radiation, sanitizers, high salt concentration;
- Relatively rapid growth rate at low temperatures.

Remember, by knowing how the organisms first get into the food, and any unusual characteristics that may permit them to multiply, one can identify potential means to control these microbes. Controls are the keys to prevention of hazards. For microorganisms, the controls simply are the means to eliminate or prevent the growth of these pathogens.

Controls are the <u>flip-side</u> of the characteristics that influence the survival of the microorganisms. For HACCP, these controls are called **Critical Control Points** (CCPs) if failure to control microorganisms at these points would allow for the potential development of a safety hazard. The most common CCPs for biological hazards include the following:

- Microbiological criteria for raw materials or ingredient (dependent on intended use, process requirements, etc.)
- Preservative factors for the food (pH, a_w, etc.)
- Time/temperature applications (cooking, freezing, holding, refrigerating, storing, etc.)
- Prevention of cross contamination
- Food handling practices
- Employee hygiene
- Packaging integrity
- Storage, distribution display practices
- Consumer directions for use (to prevent abuse)

Note that an item mentioned above would only be a CCP if failure to perform the necessary function at that point would result in a potential development of a health hazard. For example, microbiological criteria on incoming ingredients may not be a CCP, if the process of the item (e.g., thermal process) would be sufficient to eliminate all pathogenic microorganisms on the product. However, microbiological criteria would be important on an ingredient, if the ingredient were to be used in a perishable product that would receive no further processing to destroy microorganisms before consumption. Thus, high levels of a pathogen on the incoming ingredient would pose a health threat to an unsuspecting consumer.

Once a CCP has been identified, critical limits must also be established. For the example given above, limits on acceptable/unacceptable levels of microorganisms would need to be established for the incoming ingredient. For other CCPs the limits may be of a different form: time/temperature limits on heat-processed or refrigerated items, pH limits on acidified or pH-controlled foods, a_w limits on intermediate moisture foods, use of gloves or hand sanitizers with food handling practices, etc.

SUMMARY

HACCP requires an understanding of the types of microorganisms important in the safe processing of a specific food. Classification of hazardous microorganisms and indicator organisms (and related tests) is helpful for assessment of risk to establish preventative controls. The characteristics of the microorganism(s) are examined to determine the appropriate controls. Critical Control Points and their limits are then determined, and specifications for CCPs documented.

REFERENCES

Archer, D. L. 1988. The true impact of foodborne infections. *Food Technol*. 42(7):53.

Archer, D. L. and J. E. Kvenberg. 1985. Incidence and cost of foodborne and diarrheal disease in the United States. *J. Food Protect*. 48:887.

Bean, N. H. and P. M. Griffin. 1990. Foodborne disease outbreaks in the United States, 1973-1987: pathogens, vehicles and trends. *J. Food Protect*. 53:804.

Bean, N. H., P. M. Griffin, J. S. Goulding and C. B. Ivey. 1990. Foodborne disease outbreaks, 5-year summary, 1983-1987. *J. Food Protect*. 53:711.

CDC. 1985. Recommendations for preventing transmission of infection with human T-lymphotropic virus type III/lymphadenopathy-associated virus in the work place. *Morbid. Mortal. Weekly Rep*. 34(45):682.

Chung, K. C., and J. M. Goepfert. 1970. Growth of *Salmonella* at low pH. *J. Food Sci*. 35:326.

Conner, D. E., V. N. Scott and D. T. Bernard. 1990. Growth, inhibition and survival of *Listeria monocytogenes* as affected by acidic conditions. *J. Food Protect*. 53:652.

FDA. 1988. *Food Protection Unicode*. Food and Drug Administration, U.S. Department of Health and Human Services, Washington, D. C.

IFT. 1988. Virus transmission via foods (Scientific Status Summary). *Food Technol*. 42(10):241.

ICMSF. 1986. *Microorganisms in Foods 2 - Sampling for Microbiological Analysis: Principles and Specific Applications*, 2nd. Ed., The International Commission on Microbiological Specifications for Foods. Univ. Toronto Press, Toronto, Canada.

Khan, P. 1988. AIDS and the food worker. *Food Eng.* (January):79.

Levine, W. C., W. T. Stephenson, and G. F. Craun. 1990. Waterborne disease outbreaks, 1986-1988. *Morbid. Mortal. Weekly Rep.* 39/No. SS-1:1.

NAS/NRC. 1985. *An Evaluation of the Role of Microbiological Criteria For Foods and Food Ingredients.* National Academy Press, Washington, D.C.

Raatjes, G. J. M. and J. P. P. M. Smelt. 1979. *Clostridium botulinum* can grow and form toxin at pH values lower than 4.6. *Nature, London* 281:398.

Riemann, H. and F. L. Bryan. 1979. *Food-borne Infections and Intoxications*, 2nd ed. Academic Press, N.Y.

Smelt, J. P. P. M., G. J. M. Raatjes, J. S. Crowther and C. T. Verrips. 1982. Growth and toxin formation by *Clostridium botulinum* at low pH values. *J. Appl. Bacteriol.* 52:75.

Sorrells, K. M., D. C. Enigl and J. R. Hatfield. 1989. Effect of pH, acidulant, time and temperature on the growth and survival of *Listeria monocytogenes*. *J. Food Protect.* 52:571.

Tanaka, N. 1982. Toxin production by *Clostridium botulinum* in media at pH lower than 4.6. *J. Food Protect.* 45:234.

Todd, E. 1990. Epidemiology of foodborne illness: North America. *The Lancet* 336:788.

Todd, E. C. D. 1984. Economic loss resulting from microbial contamination of food. In, *Proceedings of the Second National Conference for Food Protection*, U. S. FDA, Washington, D. C.

Tompkin, R. B. and T. V. Keuper. 1982. How factors other than temperature can be used to prevent microbiological problems. In, *Microbiological Safety of Foods in Feeding Systems*, ABMPS Report No. 125, National Research Council, National Academy Press, Washington, D.C.

Young-Perkins, K. E. and R. L. Merson. 1986. *Clostridium botulinum* germination, outgrowth and toxin production below pH 4.6; Interactions between pH total, acidity and buffering capacity. *J. Food Sci.* 52:1084.

CHAPTER 5

CHEMICAL HAZARDS AND CONTROLS

by Allen M. Katsuyama

INTRODUCTION

Although biological hazards are of greatest concern because they are capable of causing extensively widespread foodborne illnesses, chemical hazards may also cause foodborne illnesses, albeit generally affecting fewer individuals. Therefore, a well designed HACCP program requires the identification of significant chemical hazards and implementation of appropriate preventive controls.

A wide variety of chemicals are routinely used in the production and processing of foods. Some examples of common types of chemicals are listed in Table 5-1. While these types of chemicals do not represent chemical hazards when used properly, some of them are capable of causing illness or even death if used improperly. Therefore, the hazard analysis must consider whether any of these chemicals is used in a manner which creates a significant food safety problem.

TABLE 5-1. CHEMICALS USED IN FOOD PROCESSING

POINT OF USE	TYPES OF CHEMICALS
Growing crops	Pesticides, herbicides, defoliants
Raising livestock	Growth hormones, antibiotics
Production	Food additives, processing aids
Plant maintenance	Lubricants, paints
Plant sanitation	Cleaners, sanitizing agents, pesticides

LAWS AND REGULATIONS

Whether or not a chemical is allowed in our food is the decision of the agency (FDA or USDA) responsible for implementing the provision of pertinent laws. Since the primary goal of such laws is to ensure that the foods available to the consuming public are safe and free of adulterants, the following acts are specifically prohibited:

1. introduction of adulterated foods into interstate commerce;
2. adulteration of foods while those foods are in interstate commerce (in transit); and,
3. receipt in interstate commerce of any adulterated food.

The Federal Food, Drug, and Cosmetic Act (FD&C Act) has several definitions for adulteration or adulterated food, two of which relate directly to chemical hazards. The FD&C Act states that "a food shall be deemed to be adulterated (1) if it bears or contains any poisonous or deleterious substance which may render it injurious to health, or (2) if it bears or contains any added poisonous or deleterious substance."

The regulations classify harmful chemicals into two major categories: (a) prohibited substances and (b) unavoidable poisonous or deleterious substances (Table 5-2). Other chemicals associated with foods are classified into the following categories according to their common uses:

- Color additives
- Direct food additives
- Indirect food additives
- Prior-sanctioned substances
- Substances generally recognized as safe
- Pesticide chemicals

CHEMICALS OF CONCERN TO THE FOOD INDUSTRY

The above regulatory classification of chemicals serves as a useful guide in addressing potential chemical hazards during the development of the HACCP program.

Prohibited Substances

The substances prohibited from direct addition to foods had been used, or had been proposed for use, for a variety of functions, including as flavoring compounds (calamus, cinnamyl anthranilate, coumarin, safrole), artificial sweeteners (cyclamates, dulcin, P-4000), preservatives (monochloroacetic acid, thiourea), a foam stabilizer (cobaltous salts), antioxidant (NDGA), or fermentation inhibitor (DEPC). The substances prohibited from uses that could lead to the indirect incorporation of the chemicals in foods were formerly used for packaging or other food-contact materials, either as adhesives or as resin components. Every company should ensure that none of these chemicals are present in ingredients and supplies that are used in the plant.

TABLE 5-2. CATEGORIES OF HARMFUL CHEMICALS

Prohibited Substances

Direct Addition

Calamus and its derivatives
Cinnamyl anthranilate
Cobaltous salts
Coumarin
Cyclamates
Nordihydroguaiaretic acid (NDGA)
Diethylpyrocarbonate (DEPC)

Dulcin
Monochloracetic acid
Chlorofluorocarbon propellants
P-4000
Safrole
Thiourea

Indirect Addition

Flectol H
Mercaptoimidazoline and 2-mercaptoimidazoline
4,4'-Methylenebis (2-chloroanaline)
Hydrogenated 4,4'-isopropylidene-diphenolphosphite
    ester resins

Unavoidable Poisonous or Deleterious Substances

Aflatoxin
Aldrin and Dieldrin
Benzene hexachloride (BHC)
Cadmium
Chlordane
Chlordecone (Kepone™)
Crotolaria seeds
DDT, DDE, and TDE
Dicofol (Kelthane™)
Dimethylnitrosamine
  (nitrosodimethylamine)
Endrin

Ethylene dibromide (EDB)
Heptachlor and Heptachlor epoxide
Lead
Lindane
Mercury
Methyl alcohol
Mirex
N-Nitrosamines
Paralytic shellfish toxin
Polychlorinated biphenyls (PCBs)
Toxaphene

Unavoidable Poisonous Or Deleterious Substances

The presence of poisonous or deleterious substances in food renders the food adulterated. However, when the presence of a poisonous substance is unavoidable -- either because the substance is necessary in the production of a food product or cannot be avoided by the use of good manufacturing practice -- FDA is authorized to establish a tolerance or action level for the substance. Action levels for those chemical listed in Table 5-2 have been established for specific commodities and are published in a "Defect Action Levels" list available from FDA. In the absence of a tolerance or action level for a specific hazardous chemical in a specific food product, none is allowed.

In all cases, foods containing levels higher than established tolerances or action levels are considered adulterated and are subject to legal action. Furthermore, blending of materials that violate a tolerance or action level with materials that are unadulterated is illegal and renders the entire blended quantity adulterated regardless of the final concentration in the blend.

The company must ensure that any poisonous substance that may be present in ingredients or food-contact materials is unavoidable and meets established tolerances or action levels. The company must further ensure that storage and handling conditions do not result in an increase in the concentration of a hazardous substance; any increase is considered to be an addition of a poisonous substance, thereby rendering the material adulterated even though the concentration may remain below an established tolerance or action level.

Color Additives

Color additives used in foods are subject to specific regulations that were developed pursuant to the color additives amendments to the FD&C Act. Except for those coloring agents listed in Table 5-3, all color additives must be certified by FDA's Color Certification Branch.

The regulations (21 *CFR* 73) describe the identity, state the specifications, list permitted uses and restrictions, and specify labeling requirements for each of the exempted color additives. The company must ensure that all requirements are consistently met when color additives exempt from certification are used.

The regulations also specify the identity, specifications, permitted uses and restrictions, and labeling requirements for each color additive subject to certification (21 *CFR* 74). Each *batch* of color additives requiring certification must be appropriately certified by FDA. The manufacturer of the color additive is responsible for obtaining certification and is prohibited from shipping any batch prior to certification. Upon receipt of certification, each container must be identified by labeling with the batch certification number.

To avoid product adulteration and potential food safety problems, food processors who use food coloring agents must ensure that each container lists the batch certification number and that each color additive is used in compliance with pertinent regulations.

TABLE 5-3. COLOR ADDITIVES EXEMPT FROM CERTIFICATION

Annatto extract
Dehydrated beets (beet powder)
Ultramarine blue
Canthaxanthin
Caramel
β-Apo-8'-carotenal
β-Carotene
Cochineal extract; carmined
Toasted partially defatted cooked
 cottonseed flour
Ferrous gluconate
Grape color extract
Grape skin extract (ecocianina)
Synthetic iron oxide

Fruit juice
Vegetable juice
Dried algae meal
Tagates (Aztec marigold) meal
 and extract
Carrot oil
Corn endosperm oil
Paprika
Paprika oleoresin
Riboflavin
Saffron
Titanium dioxide
Turmeric
Turmeric oleoresin

Direct Food Additives

A major group of chemicals used in food processing is the direct food additives category. By definition, these are chemicals that are intentionally added or incorporated directly into foods. The regulations (21 *CFR* 172) list specific "food additives permitted for direct addition to foods for human consumption" under the following functional categories:

- Food Preservatives
- Coatings, Films and Related Substances
- Special Dietary and Nutritional Additives
- Anticaking Agents
- Flavoring Agents and Related Substances
- Gums, Chewing Gum Bases and Related Substances
- Other Specific Usage Additives
- Multipurpose Additives

Additionally, regulations have also been promulgated addressing "secondary direct food additives permitted in food for human consumption" (21 *CFR* 173). These are substances that, although not necessarily added directly into food, are used in a manner or under such conditions that they may reasonably be expected to become a component of food. The approved chemicals are listed under the following categories:

- Polymer Substances and Polymer Adjuvants for Food Treatment
- Enzyme Preparations and Microorganisms
- Solvents, Lubricants, Release Agents and Related Substances
- Specific Usage Additives

In all cases, the regulations state the identity, specifications, and permitted uses and restrictions for each direct food additive.

Indirect Food Additives

Indirect food additives include chemicals that are permitted for use in food-contact materials where those chemicals might migrate from the food-contact article and become a harmful component of the food. Identities, specifications, permitted uses and limitations for specific chemicals are listed according to their following functional uses:

- Adhesives and components of coatings (21 *CFR* 175)
- Paper and paperboard components (21 *CFR* 176)
- Polymers (21 *CFR* 177)
- Adjuvants, production aids, and sanitizers (21 *CFR* 178)

Also, irradiation of foods is considered to result in a food additive and is addressed in the regulations (21 *CFR* 179).

In addition to the lubricants and sanitizers addressed in the FDA regulations, paints and other coatings used for maintaining food processing equipment and facilities must be addressed as potential indirect food additives. Although FDA only approves specific chemical compounds or chemical formulations of lubricants and sanitizers, USDA approves and publishes a list of proprietary substances, including cleaning compounds, sanitizing agents, water treatment chemicals, lubricants, and paints and other coatings.

Regardless of approval, no indirect food additive may result in imparting an odor or taste to a food product as to render it unfit for human consumption.

Prior-Sanctioned Substances

Substances that were sanctioned for use in foods by FDA or by USDA prior to September 6, 1958, the effective date of the Food Additive Amendment, are excluded from the definition of "food additive." Unfortunately, FDA does not have a comprehensive list of prior-sanctions. Those that have been identified and incorporated into the regulations (21 *CFR* 181) are limited to the following:

A. Substances employed in the manufacture of food-packaging materials -

        Antimycotics
        Antioxidants
        Driers
        Drying oils as components of finished resins
        Plasticizers
        Release agents
        Stabilizers

B. Substances used in the manufacture of paper and paperboard products -

        Acrylonitrile polymers and resins
        Sodium nitrate and potassium nitrate
        Sodium nitrite and potassium nitrite

Substances Generally Recognized As Safe

Numerous common food ingredients, such as salt, pepper, sugar, vinegar, baking powder, and monosodium glutamate are considered by FDA to be safe. Although FDA recognizes that it is impractical to list all substances that are generally recognized as safe (GRAS) for their intended use, a number of substances have been listed in the regulations (21 *CFR* 182). The GRAS substances have been listed according to the following categories:

    Spices and other natural seasoning and flavorings
    Essential oils, oleoresins (solvent-free), and natural extractives (including distillates)
    Natural extractives (solvent-free) used in conjunction with spices, seasonings, and
        flavorings
    Certain other spices, seasonings, essential oils, oleoresins, and natural extracts
    Synthetic flavoring substances and adjuvants
    Substances migrating from cotton and cotton fabrics used in dry food packaging
    Substances migrating to food from paper and paperboard products
    Adjuvants for pesticide chemicals
    Multiple purpose GRAS food substances
    Anticaking agents   Emulsifying agents
    Dietary supplements   Sequestrants
    Stabilizers   Nutrients

GRAS substances may be used in accordance with good manufacturing practice, which has been defined to include the following restrictions:

a. the quantity does not exceed the amount reasonably required to accomplish its intended physical, nutritional, or other technical effect in food;
b. the quantity that becomes a component of the food is reduced to the extent reasonably possible; and
c. the substance is of appropriate food grade and is prepared and handled as a food ingredient.

Pesticide Chemicals

The manufacturing, distribution, sale, and uses of all pesticide chemicals (insecticides, rodenticides, fungicides, herbicides, plant regulators, defoliants, desiccants, etc.) are closely regulated by the Environmental Protection Agency (EPA) under the authority of the Federal Insecticide, Fungicide, and Rodenticide Act (FIFRA) as amended. EPA approval of each pesticide formulation includes specific limitations regarding the means by which the chemical may be applied, conditions of application, permitted concentrations, the target organisms against which the chemical may be employed, use restrictions, and requirements for the disposal of the pesticide and its containers. Additionally, each agricultural pesticide is approved only for specific crops. The use of any pesticide, including those used in a plant pest control program, must comply strictly with the instructions and information on the label.

In addition to determining which pesticides may be used on agricultural crops, EPA also has the responsibility to determine tolerances or exemptions from tolerances for pesticide residues on raw agricultural commodities (40 *CFR* 180) and to establish tolerances for pesticide residues in processed food (40 *CFR* 185).

The pesticide tolerances in processed foods are enforced by FDA. Pesticide residues in packaging materials for processed foods, and pesticides used as preservatives in processed foods or as sanitizers of food-contact surfaces are considered food additives and are also regulated by FDA.

DESIGNING THE HACCP PROGRAM

Unacceptable chemical hazards may be created by the use of various chemicals at several points in the food production chain. The regulatory categories of chemicals cited above provide guidelines for identifying potentially significant chemical hazards. If such chemical hazards are present, they may be addressed at one or more of the following points:

- Prior to receipt of food ingredients and packaging materials,
- Upon receipt of these materials,
- During processing at the points where chemicals are used,
- During the storage of food ingredients, packaging materials, and hazardous chemicals,
- During the use of cleaning agents, sanitizers, lubricants, and other sanitation and maintenance chemicals, and
- Prior to the shipment of finished goods.

(Note: See Table 5-4 at the end of this chapter.)

Prior To Receipt

Suppliers may be involved in reducing the significance of chemical hazards associated with ingredients and supplies. Examples include pesticide residues on raw agricultural commodities; naturally occurring poisonous chemicals, such as mycotoxins in grains and paralytic shellfish toxins in mollusks; added poisonous chemicals, such as antibiotics and growth regulators in meat and poultry; and toxic chemical compounds in packaging materials and maintenance supplies. Although the supplier is involved in eliminating or reducing the risks associated with use of these chemicals, the HACCP system must delineate appropriate measures to insure that any remaining significant hazards are under control.

Manufacturers should develop specifications for all ingredients and packaging materials used to produce the finished product, as well as maintenance, sanitation, and other chemicals used in the plant. References to regulations or regulatory approval should be cited when pertinent. For example: "only approved pesticides may be applied to agricultural commodities" and "pesticide residues must comply with established tolerances"; food coloring agents must be FDA-certified and each container must clearly show the FDA batch certification number; cleaning chemicals "must be USDA approved" or sanitizing agents "must be approved by FDA."

A letter of guarantee should be obtained from all suppliers and vendors. The letter should state that the supplier or vendor guarantees that every item shipped to the company meets the specifications that have been provided.

Many companies certify or qualify suppliers and vendors before purchasing ingredients and supplies. The primary purpose of certification is to insure that the supplier or vendor is complying with pertinent regulatory requirements, such as the good manufacturing practice regulations, and is capable of providing items that meet specifications.

Upon Receipt

Although specifications, letters of guarantee, and vendor certifications will help insure the chemical safety of ingredients and supplies, additional measures must be taken when the materials are received at the plant. Each vehicle must be inspected before any items are unloaded. If a chemical odor or spilled substance of unknown origin is noted inside of a trailer or railcar or on a pallet or container, the shipment must either be rejected or placed on hold for further evaluation. Materials should also be inspected during the unloading process to insure that no chemical hazards are found among the individual pallets or containers.

The controls instituted prior to receipt of ingredients and supplies eliminate the need to routinely test received materials. However, a periodic sampling and testing protocol is prudent for monitoring supplier performance.

Processing

One method of controlling the chemicals used in a food processing plant is to insure that only "approved" chemicals are used at the facility. Specifications and letters of guarantee may be used for this purpose. Additionally, a knowledgeable individual should be assigned the responsibility for assuring that all chemicals received, stored, and used in the plant are approved. The processing steps where individual ingredients, processing aids, and food additives are used must be addressed during the development of the HACCP plan.

To control the in-house use of chemicals, the points of use for each chemical must be addressed. Batching sheets must be posted whenever formulated products are being manufactured. Logs should be developed for recording the usage of chemicals such as nitrites and sulfiting agents. Since employee practices, including proper storage, handling, and use of chemicals in exposed food areas, are important, all food handlers must be thoroughly trained. Unlabeled chemical containers are a serious problem in any food manufacturing facility and can only be avoided by thorough employee training and a strong company policy regarding the use and storage of chemicals in food production areas.

Regularly scheduled in-house audits should be performed to insure that hazardous chemicals are being adequately controlled in processing areas. Each audit should include observation of production practices, review of product formulations, verification of batching sheets and usage logs, where applicable, and confirmation that only approved chemicals are being used and are being stored and handled appropriately.

Storage

Cross contamination is always of concern in a warehouse if chemicals are stored in close proximity to raw ingredients, packaging materials, or finished products. All chemicals must be stored in tightly sealed containers. Hazardous or toxic chemicals must be stored in physically separated, secured enclosure accessible only to authorized personnel. Food additives and other chemicals, especially ingredients such as nitrites and sulfiting agents, must be stored to prevent cross contamination. Packaging materials in storage must be covered to protect against contamination. Common sense and strict adherence to good manufacturing practices should provide adequate control.

Sanitation and Plant Maintenance

All chemicals used in the plant sanitation and maintenance programs should be approved by the appropriate regulatory agency to avoid creating potential chemical hazards through their use. An excellent reference for such chemicals is the document entitled "List of Proprietary Substances and Non-food Compounds" published by the USDA. Every cleaner, sanitizer, lubricant, paint and coating, and other type of chemical that has been approved for use in USDA-inspected facilities is listed by manufacturer, brand name, and specific approved use. The list is recognized by FDA which also publishes an approved list of specific chemical compounds, but not brand names, of sanitizers, water treatment and other chemicals (see 21 *CFR* 178). The individual responsible for purchasing the chemicals used in the plant should refer to these lists. If a question arises regarding the acceptability of a chemical, the supplier of the chemical should be asked for a copy of a letter from a regulatory agency stating that the chemical has been approved for use in food facilities.

Chemical Residues

Misuse or negligent use of cleaning and sanitizing chemicals can create chemical hazards in foods. Detailed, written cleaning and sanitizing procedures should be developed for each piece of equipment and every line in the facility. The cleanup crew must be thoroughly trained to insure that the procedures are understood and explicitly followed. The procedures must insure that no harmful chemical residues are left on food contact surfaces, especially inside product conveying lines. A simple way to assure that previously cleaned lines are free of harmful cleaning chemicals is to test the final rinse water with pH-indicating paper.

Pesticide Usage

Even a well designed integrated pest management (IPM) system will require the occasional use of pesticide chemicals, such as fogging inside the facility with a non-residual insecticide or applying residual sprays outdoors. Whether an outside pest control operator (PCO) is contracted or the total pest management program is handled in-house, labels of all pesticide chemicals being used at the facility should be kept on file. Pesticide usage records must be maintained to show when each pesticide is used, the quantity used, and where and how the application was made.

Pesticide labels clearly state the concentration, method of application, and the target organism for each chemical. Using a pesticide in any other manner, including against a pest not identified on the label, constitutes a violation of the Federal Insecticide, Fungicide, and Rodenticide Act and could result in all foods stored in a mistreated area as being deemed "adulterated" under the FD&C Act.

When poison baits are being considered for controlling rodents at a facility, insure compliance with all regulations. While bait stations containing poisons, when properly used outdoors, will not create food safety problems, the chance for food contamination from interior bait stations is of great concern. Therefore, it is strongly recommended that no bait be placed inside a food processing plant. Although regulations vary from state to state, the GMP regulations require limiting the use of poison baits to the outside of the plant.

The storage of pesticides, as with all toxic chemicals, must be strictly controlled. The GMP regulations require that such chemicals be stored securely in an enclosed area accessible only to authorized employees. Appropriate warning signs must also be posted at these storage locations.

Prior To Shipment

All vehicles should be inspected prior to loading finished goods. Each vehicle must be free of chemical or other objectionable odors and residues of unknown materials. Although recently enacted legislation addresses chemical contamination of foods in commercial vehicles, nothing substitutes for in-house awareness through routine, careful inspection and documentation of the vehicle inspection.

TABLE 5-4. EXAMPLES OF CHEMICAL USED IN FOOD PRODUCTION, POINTS OF CONTROL, AND CONTROL MEASURES

CHEMICAL	POINT OF CONTROL	CONTROL MEASURE
Raw Materials		
Pesticides, toxins, hormones, antibiotics, hazardous chemicals	Prior to receipt	Specifications, letters of guarantee, vendor certification, approved uses.
	Upon receipt	Vehicle inspection, tests, controlled storage conditions.
Color additives, inks, indirect additives, prohibited substances in packaged ingredients and packaging materials	Prior to receipt	Specifications, letters of guarantee, vendor certification, approved uses.
	Upon receipt	Vehicle inspection, proper storage.
Processing		
Direct food additives	Prior to receipt	Review purpose, purity, formulations (quantity), label requirements.
	Point of use	Handling practices.
Color additives	Prior to receipt	Review purpose, exempted/certified, labeling requirements.
	Point of use	Handling practices, quantities used.
Water additives	Boiler/water treatment systems	Approved chemicals, handling practices, quantities used.
Building and Equipment Maintenance		
Indirect food additives, paints, coatings, lubricants	Prior to use	Specifications, letters of guarantee, approved chemicals.
	Point of use	Handling practices, quantities used, proper storage.
Sanitation		
Pesticides	Prior to use	Approved chemicals, procedures/uses.
	Point of use	Handling practices, label instructions, surfaces protected, cleaned after application.
Cleaners, sanitizers	Prior to use	Approved chemicals, procedures.
	Point of use	Procedures, adequate rinsing.

TABLE 5-4.   (Continued)

CHEMICAL	POINT OF CONTROL	CONTROL MEASURE
Storage and Shipping		
Cross contamination	Storage area	Organized by type of materials; toxic chemicals secured/limited access; inventory all chemicals.
All types of chemicals	Shipping vehicles	Inspect and clean vehicles before loading; ship food and chemicals separately.

SUMMARY

The following steps are required in developing and implementing a system for control of chemicals used in a food processing facility:

1. Use only approved chemicals. Develop specifications and obtain letters of guarantee from all suppliers of chemicals, ingredients, and packaging materials.
2. Keep an inventory of all chemicals, including food additives and coloring agents, that are used in the plant.
3. Review current procedures for using all chemicals, including product formulations.
4. Audit the use of all chemicals, including the monitoring of employee practices.
5. Institute appropriate in-house tests.
6. Assure adequate employee training.
7. Keep abreast of new regulations.

Although specifications and letters of guarantee may be kept on file, periodic tests to insure supplier performance are desirable. Visits to supplier facilities provide indications of the degree of safety afforded their products and will enhance or detract from your confidence level for each supplier.

New regulations are continually being promulgated and existing regulations are frequently revised. Therefore, be sure that someone in the organization keeps updated on regulations governing the use of chemicals to avoid any problems concerning chemical hazards.

Finally, bear in mind that if a food is chemically adulterated, whatever the cause -- by the use of an ingredient containing an illegal chemical or excessive residue, by an operating error during production, or by shipment of finished foods in a chemically tainted container or vehicle -- the responsibility will fall upon the company producing the finished product. The company must act in a responsible manner and have proper documentation of every step. A carefully conceived and well implemented system to control the use of chemicals can achieve these goals and help to assure the safety of all products.

REFERENCES

Corlett, D.A., Jr. and R.F. Stier. 1991. Risk assessment within the HACCP system. *Food Control* 2:71-72.

Environmental Protection Agency. 1992. Tolerances and exemptions from tolerances for pesticide chemicals in or on raw agricultural commodities. Title 40, *Code of Federal Regulations*, Part 180. U.S. Government Printing Office, Washington, DC. (Issued annually)

Environmental Protection Agency. 1992. Tolerances for Pesticides in Food. Title 40, *Code of Federal Regulations*, Part 185. U.S. Government Printing Office, Washington, D.C. (Issued annually)

FDA. 1992. Color additive regulations. Title 21, *Code of Federal Regulations*, Parts 70-82. U.S. Government Printing Office, Washington, DC. (Issued annually)

FDA. 1992. Food additive regulations. Title 21, *Code of Federal Regulations*, Parts 170-189. U.S. Government Printing Office, Washington, DC. (Issued annually)

FDA. 1992. Unavoidable Contaminants in Food for Human Consumption and Food-Packaging Material. Title 21, *Code of Federal Regulations*, Part 109. U.S. Government Printing Office, Washington, DC. (Issued annually)

USDA. 1992. *List of Proprietary Substances and Nonfood Compounds Authorized for Use under USDA Inspection and Grading Programs*. USDA, FSIS, Washington, DC. (Revised periodically)

CHAPTER 6

PHYSICAL HAZARDS AND CONTROLS

by Allen M. Katsuyama

INTRODUCTION

While biological and chemical hazards can present public health risks that may affect large numbers of people, physical hazards usually create problems only for an individual consumer or a small number of consumers. Physical hazards typically result in personal injuries such as a broken tooth, cut mouth, or a case of choking. Therefore, when developing a HACCP plan, consideration must be given to physical hazards and their controls.

FOOD SAFETY VS. AESTHETICS

Physical hazards are represented by foreign objects or extraneous matter that are not normally found in food, including such items as metal fragments, glass particles, wood splinters, and rock fragments or stones.

Differentiation must be made between foreign objects that are capable of physically injuring a consumer and those that are aesthetically unpleasing. Since HACCP deals solely with food safety, those physical contaminants capable of causing injuries, such as glass, metal, or objects which could cause a consumer to choke, must be addressed. Sources of such materials must be identified and stringent controls put in place at appropriate locations to protect the finished product from these types of contaminants. Whether or not such measures are a part of the HACCP plan will depend upon an evaluation of the actual risk and severity of the hazard as determined during the hazard analysis.

In some instances, physical contaminants may also include "filth," such as mold mats, insects and rodent droppings. Foreign objects are responsible for the vast majority of consumer complaints, as will undoubtedly be confirmed by a review of consumer complaint files. Thus, filth in foods is often the basis for alleged "mental anguish" and similar litigious claims. Note that although extraneous matter normally categorized as filth may not actually injure a consumer, the regulatory agencies can initiate action when it is deemed that foods are adulterated by filth, whether or not a public health threat actually exists. Thus, even though control of filth usually will not be part of a HACCP plan, firms must still comply with appropriate legal requirements.

SOURCES OF PHYSICAL HAZARDS

As with biological and chemical hazards, there are several sources of physical hazards (Table 6-1). Physical hazards in finished products may arise from several sources, such as:

- Contaminated raw materials
- Poorly designed or poorly maintained facilities and equipment
- Faulty procedures during production
- Improper employee practices

TABLE 6-1. EXAMPLES OF PHYSICAL HAZARDS AND THEIR SOURCES

PHYSICAL HAZARD	SOURCE OR CAUSE
Metal	Bolts, nuts, screws, screens/sieves, steel wool, metal fragments
Glass	Light bulbs, watch crystals, thermometers, insect bulbs
Wood splinters	Crates, pallets, equipment bracing, overhead structures
Insects	Environment, electrocuters, incoming ingredients/supplies
Hair	Meat ingredients, employees, clothing, rodents/other animals
Mold, mold mats	Poor sanitation - inadequate cleaning of equipment/facility
Rodents/droppings	Inadequate rodent controls, incoming ingredients
Gum, wrappers	Poor employee practices
Dirt, rocks	Raw materials, poor employee practices
Paint flakes	Equipment, overhead structures
Jewelry, buttons	Poor employee practices
Cigarette butts	Poor employee practices
Band-aid	Poor employee practices
Writing pen caps	Poor employee practices
Carcass ID tags	Slaughter house
Hypodermic needles	Veterinarian
Bullets/shot/BBs	Animals shot while in fields
Feathers	Poor sanitation, inadequate pest (bird) controls
Grease	Poor equipment maintenance program
Gasket materials	Inadequate equipment preventive maintenance

Raw Material Receipt

Controlling foreign objects in incoming raw materials and ingredients begins prior to receipt. Material specifications, letters of guarantee, and vendor inspection and certification will eliminate or significantly minimize foreign objects associated with received goods.

Equipment capable of detecting and/or removing potential foreign materials should be placed in-line for added protection. Some appropriate pieces of equipment are listed in Table 6-2. Proper installation, regularly scheduled maintenance, and regular calibration and inspection are essential when relying on such equipment for preventing physical hazards.

TABLE 6-2. EQUIPMENT FOR DETECTING OR REMOVING PHYSICAL HAZARDS

EQUIPMENT	FUNCTION
Magnet	Removes hazardous metallic metals
Metal detector	Detects hazardous ferrous objects larger than 1 mm and nonferrous objects larger than 2 mm
Screen or sifter	Removes foreign objects larger than size of openings (mesh)
Aspirator	Removes materials lighter than product
"Riffle board"	Removes stones from dry beans and field peas
Bone separator	Removes bone chips from meat and poultry products

Facility

Strict compliance with good manufacturing practices regulations will insure that the facility does not become a source of physical hazards in foods. Properly protected light fixtures, careful design of facilities and equipment and their adequate maintenance should prevent contaminants from the facility from becoming incorporated into product. Keeping the facility free of pests will also protect products from foreign materials of pest origin.

Processes/Procedures

Since processes and procedures are unique to each facility, a comprehensive, thorough evaluation must be made to identify hazardous practices and hazardous areas of manufacture. If a process or procedure can create a hazard, such as a bucket elevator or meat grinder in which the generation of metal fragments due to contact between equipment components is a common problem, a change in the process or procedure may be warranted. As another example, a written glass breakage policy is mandatory for all glass filling operations; the policy must include procedures for stopping the line and removing potentially affected containers whenever a breakage occurs. When such special precautions as these are necessary to the production of safe food, consideration to be given to the inclusion of these control measures in the HACCP plan. Alternatively, special precautions, such as the installation of magnets, or glass or metal detectors, must be taken to provide adequate control of potential physical hazards.

Employee Practices

Unfortunately, poor employee practices are responsible for the majority of physical contaminants entering product during production. Pull tabs from soft drink cans, bottle caps, hair pins, cigarette butts, and bandages are examples of contaminants from employees. Adhering to regulations regarding proper outer attire, hair restraints, and the absence of jewelry will help prevent many problems. Employee education and supervision are the primary control measures for these foreign materials.

While maintenance personnel play a vital role in keeping plants operating, they as a group are probably the worst offenders of good employee practices. It appears to be an industry-wide problem to get mechanics to work in a sanitary manner. Therefore, the maintenance procedures should delineate specific steps to be followed whenever there are equipment malfunctions and after routine maintenance work. The steps should include a careful inspection of the equipment and surrounding areas for loose hardware and tools, and a complete cleaning and sanitizing of the line prior to restarting the operations. Working with the maintenance department to establish an effective protocol will help the company avoid problems with physical hazards.

PHYSICAL CONTAMINANTS AND CONTROLS

Many common physical contaminants, their sources, and control measures are summarized in Table 6-3. The summary may be used as a guide to physical hazards which should be considered during the hazard analysis.

TABLE 6-3.   EXAMPLES OF PHYSICAL CONTAMINANTS, SOURCES AND CONTROL MEASURES

CONTAMINANT	SOURCES	CONTROL MEASURES
Glass	Light fixtures	Shatter-proof bulbs, shields
	Clock faces, mirrors	Replace with plastic
	Thermometers, glass containers	Glass breakage procedure
Insulating materials	Building, water and steam pipes	Inspect, maintain, use appropriate materials
Personal effects (jewelry, pens, etc)	Employees	Education, supervision
Metal fragments, nuts, bolts, screws, etc.	Ingredients	Specifications, letters of guarantee
	Machinery	Inspect, preventive maintenance
	Maintenance personnel	Education, supervision
	Processing steps	Magnets, metal detectors
	Finished product	Metal detector
Pests	Grounds	Eliminate harborages, extermination
	Building	Tight construction (exclusion), IPM
	Ingredients	Specifications, letters of guarantee, inspection, proper storage
Wood	Building	Inspection, maintenance
	Equipment/utensils	Eliminate
	Palletized goods	Inspection, clean before use
Strings, twist-ties, wires, clips	Bagged ingredients	Inspect, remove before use, sieves/screens, magnets
Stones in dry beans	Bean washing system	Stone traps ("riffle boards")
Hypodermic needles, bullets, shot, BBs	Incoming meat/poultry	Metal detector

SUMMARY

Prevention and control of physical hazards at each facility includes the following:

- Complying with good manufacturing practice regulations,
- Using appropriate specifications for ingredients and supplies,
- Obtaining letters of guarantee from all suppliers,
- Utilizing vendor certification,
- Identifying types and sources of physical hazards,
- Determining critical control points,
- Installing equipment that can detect and/or remove physical hazards,
- Monitoring the critical control points and documenting control performance, and
- Training employees.

REFERENCES

Corlett, D.A., Jr. and R.F. Stier. 1991. Risk assessment within the HACCP system. *Food Control* 2:71.

FDA. 1989. *The Food Defect Action Levels*. FDA/CFSAN. Washington, D.C. (Revised periodically)

Imholte, T. J. 1984. *Engineering for Food Safety and Sanitation*. Technical Institute for Food Safety. Crystal, MN.

Rhodehamel, E. J. 1992. Overview of biological, chemical, and physical hazards. In, Pierson, M. D. and D. A. Corlett, Jr. (eds.), *HACCP Principles and Applications*. Van Nostrand Reinhold, New York, NY.

SECTION III

DEVELOPING HACCP PLANS

Initial Tactics in Developing HACCP Plans

Hazard Analysis and Identification of CCPs

Critical Limits, Monitoring and Corrective Actions

Recordkeeping and Verification Procedures

Workshop Flow Diagrams and Forms

CHAPTER 7

INITIAL TACTICS IN DEVELOPING HACCP PLANS

by K. E. Stevenson

INTRODUCTION

The preceding chapters in this manual have described the development of the HACCP system, and provided examples of how HACCP was first implemented. In addition, the document which contained the recent revision of the HACCP principles was presented, along with detailed information on various types of hazards and controls. This chapter, and the following chapters in Section III, describe the procedures to use in applying the HACCP Principles to develop a HACCP Plan for specific products and processes.

The 1992 NACMCF document describing use of HACCP concepts and the HACCP Principles (Chapter 2) states that HACCP Plans must be specific for the product and process. Generic HACCP Plans developed for a common product or group of products may be very helpful in providing a substantial amount of material for a HACCP Plan, but each plan must be developed based upon the unique conditions which define and produce each food product. These include, but are not limited, to the following:

- Suppliers,
- Ingredient specifications
- Batches of ingredients
- Formulation
- Product specifications
- Facility and layout
- Types of equipment
- Equipment design
- Preparation procedures
- Operating conditions

- Processing parameters
- Employee practices
- Packaging materials
- Storage and warehousing
- Distribution
- Retail handling and display
- Product shelf-life
- Label
- Instructions to the consumer

Prior to development of a HACCP plan, upper management must make a commitment to support, the use of HACCP--both financially and in spirit. Once this important step has been taken, there are several steps which must follow prior to the actual application of the seven HACCP Principles.

ASSEMBLE THE HACCP TEAM

HACCP Coordinator

The first procedure in assembling a HACCP Team is to appoint a HACCP Coordinator. This individual will have overall responsibility for the development, organization and management of the HACCP program. The HACCP System is normally associated with the Quality Assurance (QA) function in a company. Thus, in most small to medium-sized companies, the HACCP Coordinator will be the Manager or a key QA employee. In larger companies, the HACCP component may be given high visibility by placing the HACCP unit under the auspices of a Vice President in charge of food safety.

Whether in a large company or a small one, the HACCP Coordinator must have the management skills and must be provided the resources necessary to implement the company's HACCP Policy and Objectives.

The HACCP Team

The HACCP Team is a multi-disciplinary unit which has the responsibility of developing HACCP Plans in accordance with the HACCP concepts and the company HACCP Policy and Objectives. This team should not be just another group of QA personnel; it should consist of personnel with skills and expertise in supervision and a wide variety of technical areas. This includes representatives from Engineering, Maintenance, Microbiology, Production, QA, Regulatory, etc.

While one or more members of the team may have extensive knowledge of, and/or experience with HACCP Systems, this is not a prerequisite for membership on the HACCP Team. In many instances, particularly with small companies, it may be necessary to obtain assistance from consultants and other outside experts in order to ensure the proper development and application of HACCP.

The HACCP Team does not need to have knowledge concerning every facet and detail of the products and processes in a facility or within a company. Ad hoc groups and project-specific teams can be utilized to provide local knowledge and expertise associated with various products and operations as they need to be addressed.

After the HACCP Team is appointed, they will begin to plan, develop and implement a HACCP Plan. Once the planning is done, they will need to gather a considerable amount of information before they can apply the HACCP Principles to the operations. This will include information on the facilities, equipment, processes, products, packaging materials, and other items and operations which may affect food safety. This background material is crucial to the development of HACCP Plans because it provides detailed information to which the HACCP Principles are applied. The remainder of this chapter provides further descriptions of the types and nature of the materials which are useful. (See Chapter 12 for additional information concerning the responsibilities and duties of the HACCP Team.)

DESCRIBE THE FOOD AND ITS INTENDED USE

Although the importance of descriptive information about the product and its intended use is often underrated, the purpose of this step is to obtain as many details concerning the product and its distribution as are practicable. Describing the product and its intended use may be accomplished by answering questions, such as:

- What is the product? (e.g., frozen fried chicken thighs; buttermilk pancake mix; chocolate ice cream)
- What is the intended use? (e.g., retail, food service, further manufacturing)
- What is the nature of the product? (e.g., fresh, canned, dried, vacuum packaged)
- What type of storage and distribution is required? (e.g., frozen, refrigerated, ambient)
- What is the shelf-life of the product?
- What preparation procedures are required? (e.g., ready-to eat, heat-and-serve, prepare and bake, reconstitution)
- What is the potential for mishandling?
- Are there any other special considerations which need to be addressed?

Note that this list seeks mainly to describe the product in consumer terms, and it is not exhaustive. Also, the questions will vary since the product description for HACCP must be tailored to the individual product. In some instances it may be necessary to include information concerning the consumers, if the product is intended for use by people who are immunocompromised. (Note: The intended use of a food should be based upon the normal use of the food by end users or the consumers.)

The next element involves preparing a detailed formula for the product. Also, it is important to develop a list of every ingredient or chemical which may find its way into the product, whether or not the items are listed on the label. This should include items such as pesticides used during growth of crops, processing aids, chemical sanitizers, etc.

The importance of such a list can be illustrated by the following example. Historically, sulfites have been used as processing aids for a variety of products and processes. In many instances, sulfites were not listed specifically on product labels since they were only used as processing aids and the levels of sulfites in the final products were quite low. When it became apparent that sulfites represented a health hazard to some individuals who were extremely sensitive, manufacturers were required to label products which contained relatively low amounts of sulfites. For manufacturers who had developed detailed knowledge of their products and ingredients, this amounted to a detailed computer search to determine what ingredients and products contained sulfites. However, for others who did not have this type of information, numerous recalls of products were initiated when low levels of sulfites were found to be present in some ingredients.

Detailed information of the formula and ingredients also alerts the HACCP Team to potential problems. Certain types of ingredients are known to be sources of specific microbiological hazards. For example, some dairy products, meat, eggs and other products of animal origin are recognized as sources of salmonellae, while cured meats and pasta may support staphylococcal growth and enterotoxin formation under certain conditions.

Nitrite in cured meats and antibiotics in milk and products of animal origin are examples of potential chemical hazards associated with specific ingredients. In many instances, the use of certain types of ingredients and equipment lead to contamination by specific foreign objects which represent physical hazards. The use of wire brushes, bucket elevators, and sifters may lead to product contamination with wire, metal components and metal shavings. Ingredients and products packaged in glass, or manufactured in a plant which packs products into jars, represent potential sources of contamination with broken glass.

The nature of the food, the shelf-life and packaging are also important to food safety. Historically, packaging has been used to protect food from adulterants and contamination. However, today's microwave-active packaging, selective barrier films, vacuum packaging, and recycled packaging materials present new and more complex food safety issues. Key issues concerning microbiological hazards associated with packaging are the potential growth and toxin production by *Clostridium botulinum* in products packaged in selective barrier films, and vacuum or modified atmosphere packaging, and the potential growth of *Listeria monocytogenes* in extended shelf-life refrigerated products.

DEVELOP A FLOW DIAGRAM WHICH DESCRIBES THE PROCESS

Development of the flow diagram should be considered a detailed compilation of material associated with the ingredients, storage, preparation, processing, packaging, storage and distribution of the product. The basic document is a simple (block) flow diagram showing the locations where specific ingredients are added in the system, the individual preparation and processing steps which occur, as well as, the associated machinery used in these operations.

The information in the flow diagram is used to evaluate whether or not hazards exist associated with the various stages depicted. Experience has shown that this document should not be comprised of engineering drawings because their level of complexity detracts from the food safety analyses. However, the type of equipment used in an operation is important, and additional information concerning the equipment, and possibly tolerances or specifications, should be available to the HACCP Team.

Mixing, cutting, conveying, chopping, grinding, and sifting or screening are examples of events which may have consequences with respect to physical hazards. Likewise, storing ingredients, cooking, pasteurizing, cooling, refrigerating, freezing and thawing are examples of events which may affect the safety of the product with respect to microbiological hazards.

Note that the flow diagram is used during the Hazard Analysis associated with Principle 1. And, Critical Control Points (CCPs) will be added to the flow diagram when they are identified using Principle 2.

VERIFY THE FLOW DIAGRAM

Once a flow diagram of the process has been prepared, it is imperative that the flow diagram be verified for accuracy and completeness by an on-site inspection of the facility, equipment and operations. In many instances, the simple step of verifying the flow diagram will identify deficiencies in the document which need to be corrected. (Note: This flow diagram, by necessity, is a dynamic document; it must be updated and modified so that it accurately reflects the current processes and operations.)

SUMMARY

Several steps must be taken prior to applying the seven HACCP Principles to a specific product and process. The initial steps include appointing a HACCP Coordinator and assembling the HACCP Team. Once in place, this group gathers necessary information by (a) describing the food, its intended use, shelf-life, distribution, consumers, etc., (b) developing a flow diagram and comprehensive information concerning preparation and processing operations, and (c) verifying, on-site, that the flow diagram is accurate and complete.

CHAPTER 8

HAZARD ANALYSIS AND IDENTIFICATION OF CRITICAL CONTROL POINTS

by Allen M. Katsuyama and K. E. Stevenson

PRINCIPLE 1:   **Conduct a Hazard Analysis. Prepare a List of Steps in the Process where Significant Hazards Occur and Describe the Preventive Measures.**

INTRODUCTION

The hazard analysis is a key element in developing a HACCP Plan. It is essential that this process be conducted in an appropriate manner since application of subsequent principles involves tasks which utilize the results of the hazard analysis. Thus, the hazard analysis represents the foundation for building a HACCP program.

One important area in conducting a hazard analysis is to assure that you are dealing with safety issues. Quality and economic issues (not involving safety) must be excluded. This is apparent from the HACCP definition of a "hazard" which states that it is "A biological, chemical, or physical property that may cause a food to be unsafe for consumption." Limiting the definition of hazards to safety issues helps to assure that the HACCP program will focus on the important safety issues associated with a product/process.

The objective of the hazard assessment is to determine which of the potential hazards associated with a food or a manufacturing process are of great enough significance that they must be addressed by the HACCP plan. The significance of the hazards must be determined based upon their likelihood of occurrence and their severity. The first step is to prepare a list of **potential** hazards which may be introduced or may increase in severity at the various steps in the manufacturing/preparation procedures. In making this evaluation, the HACCP team should consider both the likelihood of occurrence and the health consequences to consumers if the hazard is not adequately controlled. A hazard does not need to be addressed within the HACCP plan if it is not likely to occur and the severity of the hazard is not likely to result in adverse public health consequences. Hazards of low risk are typically addressed through "GMPs" and other "prerequisite" programs.

Once the significant hazards have been identified, preventive (control) measures must be delineated. If a significant hazard has been identified and no control measure is available, modifications must be made in the formulation or processes/procedures so the hazard is prevented, eliminated, or reduce to acceptable levels.

Revision of Principle 1

The statement printed at the top of this page describing Principle No. 1 is taken from a (March, 1992) revision of the "HAZARD ANALYSIS AND CRITICAL CONTROL POINT SYSTEM" developed by the HACCP Subcommittee of the National Advisory Committee on Microbiological Criteria for Foods (NACMCF, 1992). This represents a notable change in recommendations from the NACMCF. Previously, the NACMCF developed "HACCP Principles for Food Production" (NACMCF, 1990). The earlier version of Principle 1 also included an assessment of hazards, and it specifically described a two-step procedure for conducting a microbiological risk assessment. The first step in this procedure was to rank a food and its raw materials or ingredients according to six general hazard characteristics. The second step was the assignment of microbiological risk categories based upon how many of the general hazard characteristics were present. Note: A similar two-step procedure for the assessment of risks associated with chemical and physical hazards was published by Corlett and Stier (1991).

In various meetings, courses and workshops, there were questions concerning the validity of the microbiological risk assessment. While the procedure was designed to provide for a systematic evaluation of risks associated with a food and its ingredients, there were differences in the interpretation of the general hazard characteristics. Thus, the final assignment of risk categories was somewhat subjective. Furthermore, once a risk category was determined, there was no direct link to Principle 2, i.e. the actual risk category of a food or ingredient (which was determined using the microbiological risk assessment procedure) was not used in subsequent applications of the HACCP principles. As a result, the Refrigerated Foods HACCP Workshop Steering Committee recommended to USDA that Principle 1 be revised to exclude the microbiological risk assessment. The following two paragraphs provide the Steering Committee's recommendations concerning Principle 1:

> **The proper assessment of microbiological, chemical and physical hazards associated with food ingredients and food products is an objective process that requires expert judgement and detailed knowledge of the properties of the materials and manufacturing processes involved. In practice, this process necessarily involves a decision about the probability of hazards in specific situations. Expert sources of information, such as company scientists, universities and trade associations, should be asked to review Principle No. 1 hazard evaluations for accuracy.**

> **The qualitative Hazard Analysis aspect of Principle No. 1 should be emphasized using the A-F hazard questions as guidelines to help direct the thought process. The true value of the hazard analysis is to develop information on potential safety problems, not in attempting to rank their possible severity. Simple yes or no with a brief notation as to why there is or is not a potential problem is much more informative in establishing a HACCP program. The advantage resulting from a positive hazard characteristic judgement is that it will indicate the nature of the potential problem and guide the establishment of CCPs.**

Based upon the obvious need for a change in the description of the HACCP Principles, the NACMCF HACCP Subcommittee made revisions, particularly in Principles 1 and 2. They patterned their "new" principles after the approach developed by the Codex Food Hygiene Committee HACCP Working Group. Thus, the current version of Principle 1--which was presented at the beginning of this article--includes the identification of hazards, as well as the assessment of the risks and identification of preventive measures. In addition, the 1992 NACMCF HACCP document presents "Examples of Questions to be Considered in a Hazard Analysis" (Chapter 2, Appendix 2-A) as an aid to applying Principle 1. The original (two-step) hazard analysis format is included in Appendix I for comparison purposes.

While we have indicated that the two-part microbiological risk assessment is no longer used for Principle 1, many people believe that these "old" procedures, developed by the NACMCF, are useful. In particular, these procedures may be used to determine the relative risks of various ingredients, preparation and processing alternatives. This is particularly useful during the various stages in product development and may lead to alternate ingredients or processes which represent less risk. In addition, while people with experience in developing HACCP systems may not use these exact procedures, they represent a framework for conducting hazard analyses and risk assessments which may be used by others with less experience. The two-part microbiological risk assessment and an example of a form for use in conducting the microbiological risk assessment are presented in Chapter 11.

CONDUCTING A HAZARD ANALYSIS

Conducting a thorough analysis of the potential hazards associated with the production of a food commodity is the basis for developing a HACCP system. All significant hazards that can affect the safety of a food must be identified. Hazards which only affect product quality, not safety, may be addressed in a quality control plan, but quality concerns must *not* be incorporated into the HACCP system. Doing so will only encumber the HACCP system and detract from its sole objective of insuring product safety.

Emphasis in the original HACCP approach to food safety was placed on identifying microbiological hazards capable of causing foodborne illness. However, the importance of organisms other than bacteria, such as viruses and parasites, resulted in revision of the hazard analysis step to include all biological factors capable of rendering a food unsafe. Also, the potential seriousness of chemical and physical hazards was recognized and incorporated into the hazard analysis.

Biological Hazards

A biological hazard is one which, if uncontrolled, will result in foodborne illness. The primary organisms of concern are pathogenic bacteria, such as *Clostridium botulinum*, *Listeria monocytogenes*, *Salmonella* species, and *Staphylococcus aureus* (see Chapter 4). It is also important to identify potential hazards from other biological agents, such as the Hepatitis A virus, aflatoxin and other mycotoxins, fish and shellfish toxins, *Trichinella spiralis* and other parasites in meats, and histamine from microbial decomposition of scombroid fish.

As a first step, the HACCP team must identify and list the biological hazards associated with the finished product, its ingredients, handling procedures, manufacturing operations, storage, and distribution. A number of questions, which may be found in the NACMCF reports and other references, have been developed to assist in identifying biological hazards. Although the lists of questions cannot be provided here, the following outline will indicate the type of information the HACCP team must develop when conducting an analysis of potential biological hazards.

1. Review the list of ingredients used in the manufacture of the selected commodity.

   a. Are there hazardous biological agents inherently associated with any of the ingredients? (e.g., *Salmonella* in raw chicken; various pathogens in raw milk; *Listeria* in raw meat; *C. botulinum* in vegetables; histamine in scombroid fish; various pathogens and parasites in untreated water.)

   b. Are any of the ingredients or the finished product capable of supporting pathogens or otherwise susceptible to biological hazards due to contamination or mishandling?

2. Review the flow diagram for the selected product, placing emphasis on the handling procedures and manufacturing operations, as well as the storage methods and practices for the ingredients and the finished product.

   a. Are there any situations that may allow initially low levels of pathogens to multiply to hazardous numbers? (e.g., *Salmonella* in raw meat or poultry held for extended periods at room temperature.)

   b. Are there any situations where ingredients, work in process, or the finished product may become contaminated with pathogens? (e.g., *S. aureus* in fluid batter due to poor employee hygiene; pathogens from handlers of foods after a heat treatment step that is designed to destroy harmful microorganisms.)

   c. Are there any biological hazards that may be created by mishandling of the finished product in the distribution chain? (e.g., thawing of frozen entrees and holding in a thawed state for extended periods.)

3. List the significant biological hazards that have been identified during the hazard analysis and the point at which each significant hazard enters the process or may increase in severity (e.g., incoming ingredient, handling procedure, manufacturing operation, storage, distribution).

The list of significant biological hazards will depend on the nature of the finished product and its ingredients. The decision to list a specific hazard may be debatable since the risk associated with the hazard may be uncertain. The opinion of an expert microbiologist is helpful and should be obtained when compiling this list.

Chemical Hazards

Potential chemical hazards include toxic substances and any other compounds that may render a food unsafe for consumption not only by the general public, but also by the small percentage of the population that may be particularly sensitive to a specific chemical. For example, sulfiting agents--used to preserve fresh leafy vegetables, dried fruits, and wines--have caused allergenic reactions, including some fatalities, in sensitive individuals.

As in the case of biological hazards, the HACCP team must identify all significant chemical hazards associated with the production of the food commodity. The following outline will assist in identifying potential chemical hazards.

1. Review the list of raw materials, ingredients, and packaging materials that are used to manufacture the finished product.

    a. Are there any hazardous chemicals associated with the growing, harvesting, processing, or packaging of any item? (e.g., pesticide chemicals on raw agricultural commodities; growth hormones and antibiotics in meats, poultry, and milk; aflatoxin in nuts and grains; preservatives in dehydrated fruits and vegetables; coloring agents in any material; and unapproved chemicals in packaging materials.)

    b. Are all of the food additive ingredients approved for their intended uses? Do they meet purity requirements and are they being used appropriately?

    c. Are the food-contact packaging materials manufactured from approved chemicals? If the finished product is intended to be prepared in its package, such as in a microwave or conventional oven, are the packaging materials approved for such use?

    d. Are there labeling requirements associated with any food additive, such as for sulfites and some coloring agents? If so, do the product labels comply?

2. Review the flow diagram for the selected product and the selected manufacturing facility, placing emphasis on all of the equipment with food-contact surfaces. List all of the chemicals that are used in the plant for water treatment, equipment and building maintenance, cleaning and sanitizing, and pest control.

    a. Are all food-contact surfaces fabricated from corrosion-resistant materials, and are they free of toxic substances?

    b. Are all water treatment chemicals, such as boiler water additives, approved for use and used appropriately?

    c. Are only food-grade lubricants used in the plant? If non-approved lubricants are used, are they restricted to uses where there is no chance of product contamination?

    d. Are paints and other coatings on food-contact surfaces approved for such use?

    e. Are cleaning and sanitizing chemicals approved for use in food plants? Are they used appropriately?

    f. Are pesticides (insecticides, rodenticides) used for pest control in the plant? If so, are they approved for such use and are they being used appropriately?

    g. Are all hazardous chemicals handled and stored in a manner that precludes contamination of food-contact surfaces, raw materials, ingredients, packaging materials, and finished product?

3. List all of the significant chemical hazards that have been identified during the hazard analysis and the point at which each significant hazard occurs.

Significant chemical hazards are usually readily identified. However, guidance from a toxicologist or food chemist may be helpful. Regulations, such as FDA's lists of prohibited and approved food additives, and other government publications, such as USDA's list of proprietary substances approved for use in USDA-regulated plants, are useful references.

Physical Hazards

Physical hazards are represented by foreign objects which are capable of injuring the consumer. The HACCP team must identify the physical hazards associated with the finished product. The following outline will assist in identifying potential physical hazards.

1. Review the list of raw materials, ingredients, and packaging materials that are used to manufacture the finished product.

   a. Are there foreign objects capable of causing injury associated with any of the raw materials or ingredients? (e.g., stones in dry beans and field peas; metal clips on sausage casings; wood splinters in palletized materials.)
   b. Are there physical hazards associated with any packaging material? (e.g., glass fragments in empty jars and metal slivers in empty cans.)

2. While referring to the flow diagram for the finished product at the selected plant, inspect the physical facilities.

   a. Are there environmental sources of physical hazards in and around food storage and processing areas? (e.g., unprotected light fixtures; loose nuts, bolts, screws, or other fasteners on overhead structures; exposed or deteriorating insulation on pipes; corroded metal fixtures, such as support structures and louvers on ventilation ducts; wire, tape, twine and other impermanent materials used for "temporary" repairs.)
   b. Is any equipment capable of generating physical hazards? (e.g., splinters from wooden materials, including pallets; nuts, bolts, screws, or rivets; metal fragments from metal-to-metal contact, such as in screw conveyors and bucket elevators; glass fragments from unprotected thermometers and gauges.)
   c. Are there tools, utensils, and other implements used on or near the lines where there is a likelihood that they may fall into equipment or exposed foods? (e.g., small wrenches, sampling devices, writing implements, thermometers, gauges.)

3. List all of the significant physical hazards that have been identified and the point at which each hazard occurs.

RISK ASSESSMENT

The concept of risk assessment is important following a hazard analysis. Since numerous potential hazards may be identified for a food commodity, its ingredients and/or its processes, the likelihood of each hazard affecting the food must be ascertained. If the risk of the hazard creating a food safety problem in the food is low, or if the hazard is unlikely to occur, no further consideration of that hazard is necessary. For example, spores of *C. botulinum*, the microorganism responsible for botulism, may be present on raw tomatoes, but the natural acidic pH of tomatoes prevents the growth of the organism in canned tomatoes and, hence, the organism is not a biological hazard for that commodity. On the other hand, *C. botulinum* on tomatoes that are used in a product with a high final pH must be of utmost concern.

Knowledge of the methods used to grow and/or process each raw material or other ingredient used in the finished product is essential for an accurate assessment of the risks associated with a hazard. For example, pesticide usage information must be obtained to evaluate chemical hazards associated with raw agricultural commodities. If a processing step is employed to eliminate or reduce a hazard to a low risk, such as the destruction of *Salmonella* in spices by irradiation or treatment with ethylene oxide, that information is essential for proper identification of hazards and assessment of the risks.

PREVENTIVE MEASURES

When the hazard analysis has been completed and all significant biological, chemical, and physical hazards have been listed with their points of occurrence, the HACCP team must identify measures to prevent these specific hazards from compromising the safety of the finished product. This information will be used in Principle 2, to identify Critical Control Points (CCPs).

A review of the identified hazards usually reveals that there are preventive measures which the processor can use to control these hazards. For example, biological hazards due to pathogenic microorganisms can be controlled by heating the food to a specific minimum temperature for a specified period of time. Chemical and physical hazards can often be prevented through the use of appropriate materials and procedures. The results of the hazard analysis and identification of preventive measures should be compiled in tabular form. Table 8-1 provides some examples of "points of occurrence," identified hazards and preventive measures.

Table 8-1. EXAMPLES OF RESULTS OF A HAZARD ANALYSIS AND IDENTIFICATION OF PREVENTIVE MEASURES.

Point of Occurrence	Identified Hazard	Preventive Measures
Raw milk	Various pathogens	Pasteurization
Lake water intake	*Giardia*	Disinfection
Canned vegetables	*C. botulinum*	Adequate thermal process, container integrity
Batching of acidified foods	*C. botulinum*	Proper acidification
Various equipment	Metal fragments	Metal detectors, magnets, preventive maintenance
Product packed in jars or bottles	Glass	Container handling procedures, glass breakage policy
Fermenting sausages	*S. aureus* enterotoxin	Proper fermentation
Raw scombroid fish	Histamine	Proper cooling, temperature control

Principle 2:   Identify the Critical Control Points (CCPs) in the Process.

The HACCP Team must now identify CCPs based upon all of the significant hazards that were identified and enumerated during the hazard analysis. Using the list of preventive measures for each hazard, the HACCP Team must identify the points at which the preventive measures can be applied. Each of these points is then assessed, and the CCP is selected, if appropriate, for each hazard.

CONTROL POINTS AND CRITICAL CONTROL POINTS

Control points (CPs) and CCPs can be differentiated based upon the following definitions developed by the NACMCF:

Control Point: Any point, step, or procedure at which biological, physical, or chemical factors can be controlled.

Critical Control Point: A point, step, or procedure at which control can be applied and a food safety hazard can be prevented, eliminated, or reduced to acceptable levels.

There can be several points in a food processing system where biological, chemical or physical hazards can be controlled. However, there are likely to be only a few points, CCPs, where the loss of control will result in the production of a potentially unsafe food.

For example, metal fragments originating from processing equipment can be minimized in finished product by regularly scheduled preventive maintenance and inspection of equipment, by the installation of in-line magnets immediately after each unit that can generate metal fragments, and by the use of in-line metal detectors to check the finished product after packaging or just prior to being packaged. Although inspection, preventive maintenance, and magnets are important CPs, they alone cannot insure that the finished product is free of the physical hazard. Only the metal detector is capable of doing so. Therefore, if the presence of metal fragments is identified as a significant hazard, the metal detector is the CCP.

THE CCP DECISION TREE

To assist in identifying CCPs, the NACMCF recommends use of a CCP Decision Tree (Figure 8-1). Since publication of the NACMCF version in 1992, improvements have been proposed for the CCP Decision Tree. The version in figure 8-2 proposed recently by Dr. R. Bruce Tompkin, Vice President, Product Safety, Armour Swift-Eckrich is being considered by the NACMCF as a replacement for the original Decision Tree. The HACCP team should utilize the decision tree to evaluate each of the points where significant hazards can be prevented, eliminated, or reduced to acceptable levels. Each point should then be categorized as either a CP or a CCP. The results of this evaluation should be summarized and added to the material developed in Principle 1 (see Table 8-2).

Figure 8-1. The NACMCF (1992) CCP decision tree. (Apply at each point where an identified hazard can be controlled.)

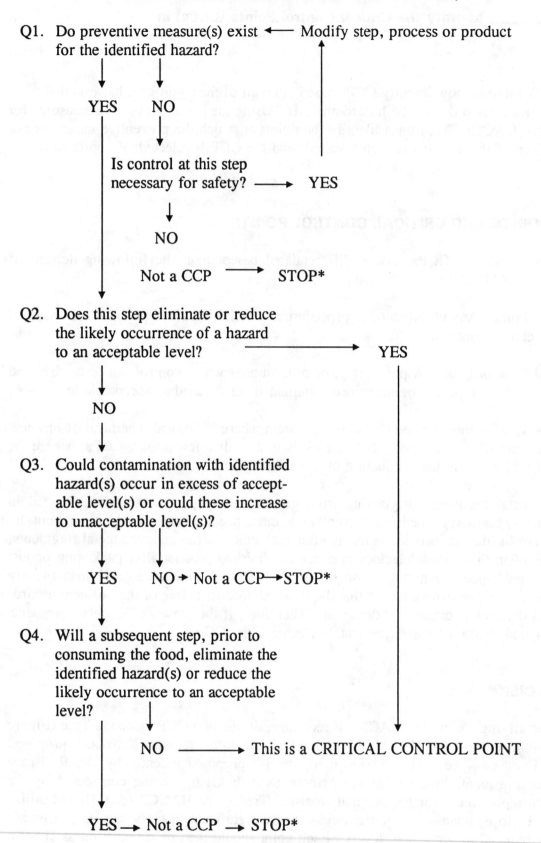

* Proceed to the next step in the selected process

Figure 8-2 The Modified CCP decision tree. (Apply at each point where an identified hazard can be controlled)

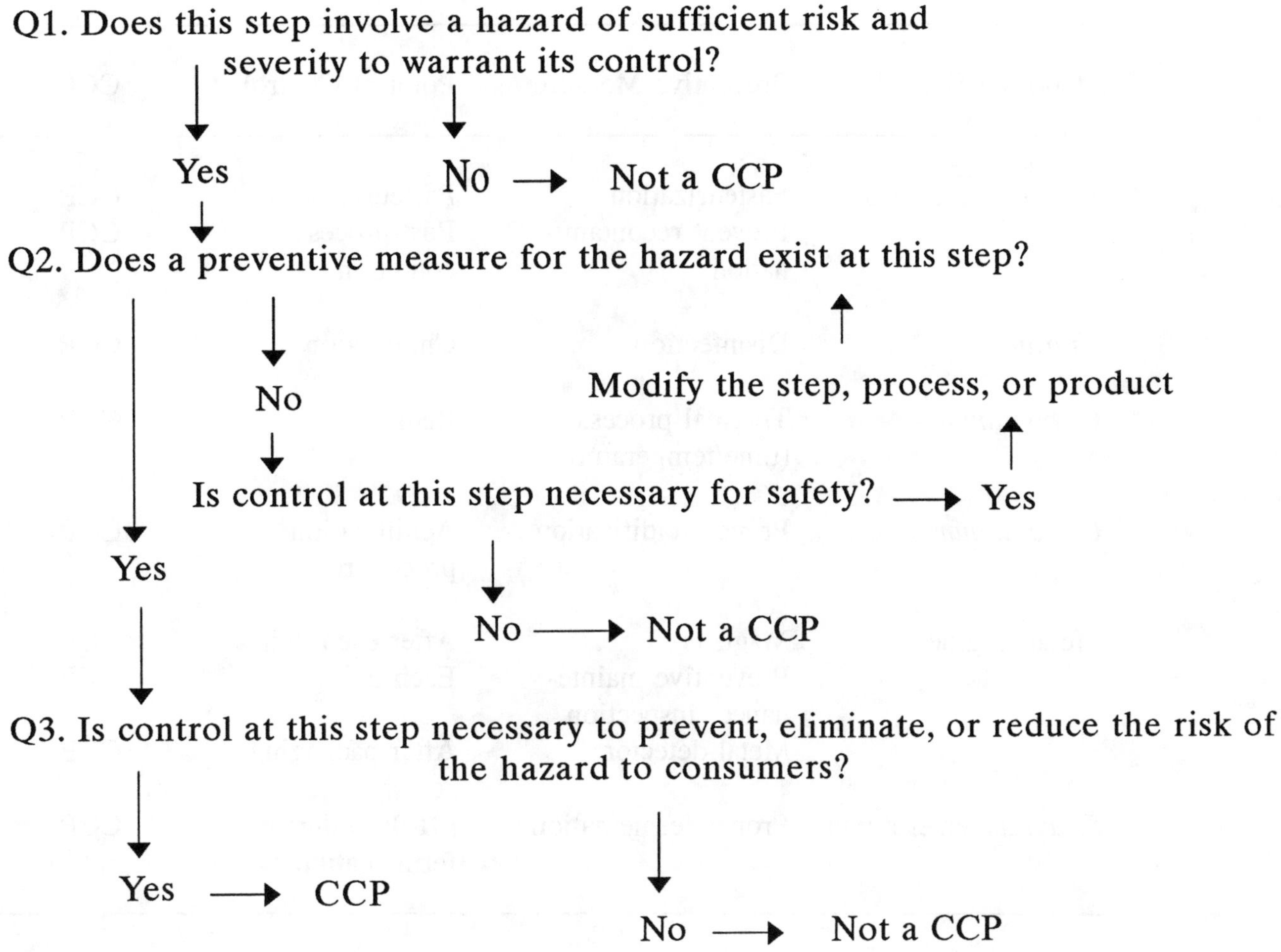

Q1. Does this step involve a hazard of sufficient risk and severity to warrant its control?

Yes

No → Not a CCP

Q2. Does a preventive measure for the hazard exist at this step?

No

Modify the step, process, or product

Is control at this step necessary for safety? → Yes

Yes

No → Not a CCP

Q3. Is control at this step necessary to prevent, eliminate, or reduce the risk of the hazard to consumers?

Yes → CCP

No → Not a CCP

Important considerations when using the decision tree:

The decision tree is used after the hazard analysis.

The decision tree then is used at the steps where a "significant hazard" has been identified. These are hazards that may reasonably be expected to occur. Non-significant hazards (i.e., of low risk and unlikely to occur) have been excluded.

A process which does not have a significant hazard does not need a HACCP plan.

Each step which is a CCP must agree with the definition: a point, step, or procedure at which control can be applied and a food safety hazard can be prevented, eliminated, or reduced to acceptable levels.

A subsequent step in the process under your control may be more effective for controlling a hazard and may be the preferred CCP.

More than one step in a process may be involved in controlling a hazard.

More than one hazard may be controlled by a specific preventative measure.

Table 8-2. EXAMPLES OF PREVENTIVE MEASURES FOR CONTROLLING IDENTIFIED HAZARDS.

Identified Hazard	Preventive Measure	Point of Control	CCP/CP
Various pathogens	Pasteurization	Pasteurizer	CCP
	Prevent recontamination	Post-process condition	CCP
Giardia	Disinfection	Chlorinator	CCP
C. botulinum	Thermal process (time/temperature)	Retort	CCP
C. botulinum	Proper acidification	Acidification procedure	CCP
Metal fragments	Magnets	After each unit	CP
	Preventive maintenance, inspection	Each unit	CP
	Metal detector	After packaging	CCP
S. aureus enterotoxin	Proper fermentation	pH drop during fermentation	CCP

CONTROL PARAMETERS AND CONTROL METHODS

A specific parameter or a combination of parameters must be controlled at each identified CCP to insure food safety. Therefore, for each CCP, the HACCP team should now list the parameter or combination of parameters and the means by which control will be applied. Examples of some preventive measures, their control parameters, and the methods or means of control are listed in Table 8-3.

Note: The CCPs that are identified serve as the basis for the HACCP system. The CCPs can be sequentially numbered for convenience (e.g., CCP #1, CCP #2, etc.). In some instances, companies prefer to number CCPs sequentially within hazard categories (e.g., CCP P1, CCP C1, for the first CCP addressing a physical hazard and a chemical hazard, respectively). While these numbering systems are primarily beneficial, in some cases they may cause confusion (when CCPs are added or deleted due to changes in the specifications, ingredients or operations).

Table 8-3. EXAMPLES OF PREVENTIVE MEASURES, CONTROL PARAMETERS, AND MEANS OF CONTROL

Preventive Measures	Control Parameters	Control Methods
Acidification	pH ≤ 4.6	pH; conc. & % titratable acidity of acidulant; ratio of solids to liquids; equilibrium pH.
pH control during fermentation	Meets pH specification within designated time	pH measurement at appropriate time intervals to comply with guidelines.
Any thermal process (pasteurization, retorting, cooking, etc.)	Time-temperature of the thermal process	Automatic temperature controller; control of product residence time; failure alarms.
Container integrity testing	Meets minimum specifications	Frequent visual examinations; regularly scheduled testing.
Metal detector	Detects minimum-size fragment	Regular calibration.

REFERENCES

Bryan, F.I. 1990. Application of HACCP to ready-to-eat chilled foods. *Food Technol.* *45*(7):70.

Corlett, D.A., Jr. and R.F. Stier. 1991. Risk assessment within the HACCP system. *Food Control* 2:71.

Ellinger, R.H. (ed.). 1990. *Total quality systems handbook - HACCP.* American Butter Institute/National Cheese Institute, Washington, DC.

IAMFES Committee on Communicable Diseases Affecting Man. 1991. *Procedures to implement the hazard analysis and critical control point system (HACCP) manual.* International Association of Milk, Food, and Environmental Sanitarians, Ames, IA.

ICMSF. 1988. *Microorganisms in foods. 4. Application of the hazard analysis critical control point (HACCP) system to ensure microbiological safety and quality.* Blackwell Scientific Publications, Boston.

NAS. 1985. *An evaluation of the role of microbiological criteria for foods and food ingredients.* National Academy Press, Washington, DC.

NACMCF. 1989. *HACCP principles for food production.* USDA, FSIS, Washington, DC.

NACMCF. 1992. *Hazard analysis and critical control point system.* Int. J. Food Microbiol. 16:1.

Pierson, M.D. and D.A. Corlett, Jr. (eds.). 1992. *HACCP: Principles and applications.* Van Nostrand Reinhold, New York, NY.

Pillsbury Company. 1973. *Food safety through the hazard analysis and critical control point system.* Contract No. FDA 72-59. Research and Development Dept., The Pillsbury Company, Minneapolis, MN.

Sperber, W.H. 1991. The modern HACCP system. *Food Technol.* *45*(6):116.

Tompkin, R.B. 1990. The use of HACCP in the production of meat and poultry products. *J. Food Protect.* 53:795.

U.S.D.A. 1992. *List of proprietary substances and nonfood compounds authorized for use under USDA inspection and grading programs.* USDA, FSIS, Washington, DC.

CHAPTER 9

CRITICAL LIMITS, MONITORING AND CORRECTIVE ACTIONS

by K. E. Stevenson, Bonnie J. Humm and Dane T. Bernard

INTRODUCTION

To this point, considerable time has been spent on the hazard analysis and identification of CCPs. Potential hazards--microbiological, chemical and physical--have been identified and their risks evaluated. Preventive measures have been identified which can control these hazards. In addition, points in the process have been identified where these preventive measures can be used to eliminate or reduce the risk of the hazards to an acceptable level. While these critical control points (CCPs) are the heart of the HACCP system, there is considerably more work to be done in developing the HACCP Plan.

The purpose of this section is to focus on what needs to be done once the CCPs have been identified. It is not enough to know that a particular hazard exists in association with the process or product. Other HACCP-related factors must be determined, including

- the critical limits (CLs) to differentiate between acceptable and unacceptable control at the CCP,
- the best method to monitor the CCP,
- the monitoring frequency, and
- action to be taken in the event of a loss of control.

CRITICAL CONTROL POINTS AND CRITICAL LIMITS

There are some general rules for CCPs and CLs with regard to monitoring and decision making.

1. The decision concerning whether or not a point is a control point (CP) or a CCP is based on the likely occurrence (risk) and severity of a hazard.

2. Appropriate CCPs must be identified *before* establishing CLs and monitoring programs.

3. The "control" exerted at most CCPs is straight-forward, and the CLs and monitoring procedures are designed to measure if the CCP is *IN* or OUT of control.

PRINCIPLE 3: Establish Critical Limits for Preventive Measures Associated with Each CCP.

INTRODUCTION

When a CCP is identified, a CL is then established which signifies whether a CCP is "in" or "out" of control. Exceeding this CL should indicate that the CCP is out of control, and therefore a potential for the development of a health hazard exists. For HACCP purposes, the *decision criteria* on this CL should be based on information relative to the following considerations:

Exceeding the critical limit indicates:

- Evidence of the existence of a direct health hazard (e.g., underprocessing of a retorted, low acid food).
- Evidence that a direct health hazard could develop (e.g., cold storage maximum temperature violated on non-cooked, refrigerated product).
- Indications that a product was not produced under conditions assuring safety (e.g., metal detector kick-outs,).
- Indications that a raw material may affect the safety of the product (e.g., audit detects staphylococcal thermonuclease at high levels).

These decision criteria reflect potential hazards. Direct microbiological hazards would indicate pathogenic microorganisms, which are all considered *direct* health hazards. Direct chemical and physical hazards typically are related to the type of food and its processing situation. A direct physical hazard may be sharp metal fragments from the malfunctioning of equipment in a processed food facility. An example of a direct chemical hazard is the presence of histamine in certain species of fish.

SETTING CRITICAL LIMITS

In many instances when microbiological CCPs are identified, the appropriate limit will not be readily apparent. A microbiologist may be able to provide a conservative recommendation on a limit that will protect the product. However, research may be necessary to further refine this limit. For example, a cold holding temperature may be identified as a CCP for a perishable ingredient on a refrigerated food processing line. The limit temperature and time at which this perishable ingredient can be held before a food poisoning hazard develops would need to be determined. Without any research, a microbiologist may recommend that the limit is 50°F for 2 hours; the ingredient and any product made if the limit were exceeded would need to be destroyed. In reality, the true temperature/time limit would not be one point but many points on a curve. Ingredients at low temperatures could be held longer, while with higher temperatures, the holding time would grow shorter. Research would be able to identify the appropriate temperature/time relationship for the ingredient, thereby providing the manufacturer with more flexibility in his schedule while still producing a safe product.

Another important factor to remember in determining the limit for CCPs is the expected variations encountered during operation of processing equipment. The best approach for assessment of this variation is to test the equipment capability before use. As an example, let's assume the minimum patty temperature necessary for destroying *pathogens* in cooked, ready-to-eat meat patties is 151°F for 41 seconds. If the cooker can control pattie temperatures only ± 5°F, the operating limits for the pattie's internal temperature would need to take this into account. In this case, by setting the target temperature too low, the ± 5°F variation may allow survival of pathogens in "cooked" patties.

Thus, the adjusted limits would need to reflect this variation:

- the minimum pattie temperature = 151°F
- the target pattie temperature = 156°F
- the maximum pattie temperature = 161°F

The CLs are:

- Internal pattie temperature, with a 151°F minimum as a CL (Note: Cooker temperature may be monitored, provided the correlation to internal temperature has been established.)
- Time at 151°F, with a 41 second minimum CL (Note: Cooker conveyor belt speed monitored, provided that the maximum speed has been determined which gives at least a 41 second residence time at the desired temperature.)
- Patties not to exceed a standard thickness (Note: Ingredient specification, or roll bar in place to assure that thickness cannot be exceeded.)

Ideally, the CL would be correlated to cooker operating parameters (cooker temperature and the conveyor belt speed). The preventive measure and CL for pattie thickness can be something as simple as using a height bar under which all patties must pass. Both the CLs for temperature and the belt speed would need to take in to account the types of monitoring devices used, their accuracy and variability, and needs for calibration.

Remember, CLs should only be used to set critical criteria at points in a process where lack of control will likely result in the development of a potential safety hazard, i.e., CCPs. They should not be set in attempts to control non-hazardous situations which are of non-safety regulatory, consumer, or economic consequence. Too many CLs, including CLs for a number of non-hazardous control points (CPs), will dilute out the HACCP effort, and result in inappropriate or inadequate monitoring and control. The end result of improper setting of CLs for CPs and the resultant excessive monitoring requirements could mean that a product with a potential safety hazard could be released to consumers. The control limits and associated monitoring procedures for CPs should be part of a standard quality program.

PRINCIPLE 4: **Establish CCP Monitoring Requirements. Establish Procedures for Using the Results to Adjust the Process and Maintain Control.**

INTRODUCTION

The types of activity that will be employed to monitor CCPs can be grouped into either *continuous* or *discontinuous* inspection. Ideally, HACCP CCPs should be monitored via continuous inspection; however, from a practical standpoint, this is not always possible.

MONITORING AND SAMPLING

Continuous Inspection

Some forms of monitoring activity are more amenable to continuous inspection than others. Continuous inspection using automated equipment, sensors, or supervision can be used to monitor CCPs with the following concerns:

- Temperature
- Time
- pH
- Moisture

Spot checks of continuous inspection activities can be useful to verify that a CCP is still in control. If a potential problem has developed at a CCP (e.g., temperature drop), once the system is brought back into control, spot checks using microbiological, chemical or physical tests may be advisable. Pre-certified ingredients (Certificate of Guarantee) can be spot checked to assure supplier compliance with specifications.

Discontinuous Inspection/Attribute Sampling

Discontinuous inspection, also called attribute sampling, is basically used under two circumstances:

- To test ingredients where microbiological, chemical or physical conditions are unknown or require testing for approval prior to processing.
- To troubleshoot when a CCP exceeds its limit (out-of-control) or when the product safety is in question. Generally, lot sampling may be required for food materials or products placed on "hold" to assess safety.

Attribute sampling can be used for chemical and physical testing. A sampling plan, based on a statistical sampling, is determined, and the units are tested for conformance. Attribute sampling for microbiological testing of hazardous microorganisms is not as well understood. A discussion of this type of testing can be found in the ICMSF book, "Microorganisms in Foods 2".

Generally, a microbiological criterion may be useful for foods or ingredients which are used or produced. A microbiological criterion should include the following:

- A statement describing the identity of the food or food ingredient.
- A statement noting the contaminant of concern, i.e. the microorganism or group of microorganisms and/or its toxin.
- The analytical method to be used for the detection or enumeration of the contaminant of concern.
- The sampling plan.
- The microbiological limits considered appropriate to the food, and commensurate with the sampling plan.

Continuous Inspection vs. Attribute Sampling

Application examples of continuous inspection versus attribute sample will help reinforce the key differences. Ideally, CCPs should be subject to continuous inspection. However, the type of processing technology being employed may affect whether continuous inspection or attribute sampling is used. For example, continuous monitoring may be used for almost all CCPs, except those that require sampling of ingredients, or in some cases, in-process products or finished products.

For some processed refrigerated products, the primary concern is for incoming ingredients containing *Salmonella* spp., *Listeria monocytogenes*, *Staphylococcus aureus* and nonpathogenic *Escherichia coli* (indicators). Each of these microorganisms can be classified according to the health hazard it presents. While this won't change, the processing or handling conditions being used will potentially increase or decrease the level of risk the microbes represent. This will then change the stringency of sampling required. A qualified microbiologist should be consulted to determine the limits for hazardous microorganisms in the various ingredients and food products.

As a practical matter, a firm may wish to impose supplier criterion such as zero <u>Salmonella</u> in ten 25-gram samples of an ingredient and expect that the supplier will provide a certificate of analysis (C.O.A.) for each lot. The receiver will then monitor product at receipt for the appropriate COA and the firm will sample occasionally to verify (principle 7) that the supplier is providing product which complies with the criteria.

Continuous inspection of a CCP provides assurance that all products produced have met the criteria for acceptability. However, this is not the case with attribute sampling Plans. With such plans, a statistical sampling of the lot is analyzed for a defect (e.g. microbiological hazard). The probability of detecting such a defect is directly related to the level of that defect in a sampled lot. With the same sampling plan, the probability of detecting a defect present at a rate of 10% of the lot is much higher than detecting a defect present at a rate of 0.1% of the lot.

When present, most microbiological hazards are present at relatively low defect levels. Thus, when dealing with moderate hazards (e.g., _Salmonella_ spp.), the probability of detecting such hazards is quite low. In fact, the probability of _accepting_ a lot with the hazardous microorganisms is quite high (often 98 to 99%). Therefore, note with caution that attribute sampling of ingredients or products provides only limited assurance that hazards have been detected (unless very large-scale sampling is employed). Whenever possible, continuous monitoring of CCPs should be used to assure product safety.

SUMMARY

HACCP systems are preventative programs where the risks of all parts of a food manufacturing sequence are assessed. Microbiological, chemical and physical CCPs are identified, limits established, and the appropriate monitoring procedure defined. The CCPs represent the control elements of a manufacturing sequence and are focused on safety. Ideally, continuous monitoring is used to assure CCP's are within limits. However, attribute sampling of critical raw materials may also be used. Regardless of which monitoring procedure is used, each CCP should have defined

- the **best** monitoring procedure
- the **frequency** of monitoring and the documentation, and
- the **decision criteria** needed to differentiate between acceptable and unacceptable control.

PRINCIPLE 5: Establish Corrective Action to be Taken When Monitoring Indicates There is a Deviation From an Established Limit.

INTRODUCTION

Since a deviation in a CCP will result in an actual or potential hazard to the consumer, appropriate action must be taken to eliminate that hazard. The specific corrective action depends on the process parameters in use and the type of food being manufactured. Due to the diversities in possible deviations, corrective actions must be developed for each CCP when they are identified and the limits and monitoring parameters are set. HACCP requires that whenever a deviation occurs, immediate corrective action is already assigned and the CCP will be brought back into control before production continues. Each deviation procedure must be outlined in the HACCP program prior to the HACCP plan being approved.

In most cases the plant will have to place the product in question on hold pending a thorough investigation of the problem. This investigation may require record review or analyses. Where the safety of the product is in question in USDA-inspected plants, testing procedures and final disposition of product must be shared with and agreed upon by the USDA inspector in charge. All product deviations at CCPs must be recorded and should remain on file for the term mandated by applicable regulations or the HACCP plan. These deviation records become an integral part of the HACCP program and must be made available to the authorities for review.

Information gathered through monitoring at each assigned CCP must be able to immediately detect potential deviations, thereby enabling the operator to take make adjustments *before* there is a need to reject or place product on hold. Remember that the HACCP plan is a "preventive program". One goal is to alert the operator of a potential problem in time for the operator to take action in order to avoid rejection of the product. Unfortunately, due to the complexity of some of our systems, complete control is not possible, and some problems can only be minimized.

CORRECTIVE ACTIONS

The following provides different options when dealing with potential or actual deviations:

1. Immediately adjust the process and keep the product in compliance within the critical limits. In this case the action is immediate, and no product is placed on hold because there has been no deviation.
2. Stop the line. Hold all product not in compliance. Correct the problem on the line, and then continue with production. Although this is a less desirable solution, it is often the scenario in food manufacturing.
3. If the deviation is the result of a problem in line design or equipment malfunction, a quick-fix may be applied in order to continue running, but a long term solution must be sought. Non-compliant product must be placed on hold. The re-evaluation process also becomes part of the HACCP program as the system evolves.

If the CCPs have been carefully identified and monitoring programs are designed for ultimate control, then actions such as number 3 above can be kept to a minimum.

Adjusting the Process

Some deviations can be controlled automatically through the use of flow diversion valves which are designed to divert product when the temperature of the product drops below a minimum set criterion. Examples of where this type of control is exerted is in a pasteurization system for milk or a hot-fill-hold system for filling tomato products.

When automatic control is not feasible, an operator can intercede and can take corrective action through the decision process outlined in the HACCP program. Whether or not the product in question can be "saved" by this method of intervention, depends on the product and the process. For example, a batch system for cooking chicken breasts could be adjusted to increase the dwell time and still reach the minimum internal temperature (IT) needed for microbial safety. If the system is a straight flow oven where the time factor cannot be changed or the temperature cannot be increased, then product exiting the oven which did not receive the minimum thermal process must be placed on hold or immediately reprocessed. Whenever possible, "corrective action" should be designed into the product line (and the HACCP system) so that almost all product deviations can be corrected in-line without having to hold product.

Other Examples of Corrective Actions:

1. Control all time/temperature dependent operations by adjusting either of the two variables while the line is still running.
2. Reroute ingredients not meeting specific criteria to another process line where the criteria are not crucial to the final safety of the product. For example, when incoming ingredients do not meet the microbiological specifications required for use in a refrigerated product, they may be sent to the canning lines where the final products receive a terminal heat process.
3. If a screw is noted to be missing on a piece of filling equipment, run suspect product through equipment which will detect any metal contaminants in the finished product containers.

Review of Records to Identify Trends

Keeping a close eye on the data from monitoring records can alert supervisors of possible equipment malfunctions or even operator errors. In this way, corrective action can be made ahead of time and a deviation at a CCP averted. Since all HACCP records should be reviewed on a daily basis, slight trends may not be as easy to identify. All critical criteria must be compared to the original limits rather than compared to yesterday's data.

Responsibility for Decision Making

It is crucial to the success of the HACCP system that responsibility for decision making is clearly delineated early on in the assignment of monitoring responsibilities. An individual knowledgeable in CCP control must have the authority to make quick decisions on the production floor in order to maintain control of a line operation.

Whenever a deviation occurs and corrective action is taken, the individual responsible for the action must record on the CCP data sheet what action was taken and by whom. If product is placed on HOLD due to the deviation the number of the hold can be written on the data sheet for easy reference.

Since any deviation in a CCP is of a safety concern rather than quality, proper documentation is critical. In USDA facilities, the inspector in charge will have to be informed of all holds, and final disposition of product in question must be agreed upon by both parties. The following questions should be asked regarding product held for deviations at CCPs.

1. What tests can be made to verify the safety of the product in question?
2. Does review of the data show the safety of the product is in serious question?
3. Can this product be diverted for use in another product where safety is assured?
4. Can the product be reprocessed or reworked in a manner resulting in 100% assurance of safety? (Example: Sending product held for possible metal contamination due to a malfunction in a metal detector through a properly operating metal detector.)
5. If the product cannot be reused, what method should be used to discard or destroy the product?
   a. Send to animal feed (inedible/unfit for human consumption).
   b. Bury in a land fill.
   c. Incinerate the product.
6. What records must be filled out and what HACCP forms should be maintained?

The disposition of all product held for deviations at CCPs must be adequately documented listing the reasons for the hold, the reasoning behind the disposition decision, the number of units and codes of all product in question, and the method of disposal. These records may be required to be made available to the regulatory agents during any HACCP verification audit.

SUMMARY

In some cases final disposition of a product may require the expertise of industry experts in toxicology, microbiology, thermal processing, or a related field in order to make a sound decision. Recommendations from such authorities should also be part of the HACCP records dealing with the deviation.

The HACCP records for deviations should include the following:

1. The actual production records or a reference to the production records relating to any products placed on HOLD. **Note:** Those deviations immediately corrected on the line which did not result in product being held should be noted on the production records in case a question ever arises regarding the safety of these products.
2. A standard form listing the following: Hold number, deviation, reason for hold, number of containers held, date of hold, date and code of product held, disposition and/or release forms, name of individual responsible for decision on disposition.
3. Recommendations for outside authorities regarding final disposition.
4. Regulatory forms regarding final disposition.
5. An accurate accounting of all units in question.
6. A statement of the standard operating procedure for handling CCP deviations.

If a HACCP plan is properly designed, all deviations will be discovered and corrected before any product leaves the facility. This is why record review is so important to the success of the HACCP program. Built into this system is a "self-correcting" feature where holes in the safety net are repaired by modifying the monitoring programs at the CCPs, thereby preventing future problems and assuring the safety of the final product.

REFERENCES

Corlett, D.A., Jr. 1991. Monitoring a HACCP system. *Cereal Foods World* 36:33.

Hudak-Roos, M. and E. S. Garrett. 1992. Chapter 7. Monitoring critical control point critical limits. In, Pierson, M. D. and D. A. Corlett, Jr. (eds.), *HACCP Principles and Applications*. Van Nostrand Reinhold, New York.

ICMSF. 1986. *Microorganisms in Foods. 2. Sampling for Microbiological Analysis: Principles and Specific Applications*, 2nd Ed., The International Commission on Microbiological Specifications for Foods. Univ. Toronto Press, Toronto, Canada.

ICMSF. 1988. *Microorganisms in Foods. 4. Application of the Hazard Analysis Critical Control Point (HACCP) system to ensure microbiological quality and safety*. Blackwell Scientific Publications, Boston.

Moberg, L. J. 1992. Chapter 6. Establishing critical limits for critical control points. In, Pierson, M. D. and D. A. Corlett, Jr. (eds.), *HACCP Principles and Applications*. Van Nostrand Reinhold, New York.

NAS. 1985. *An Evaluation of the Role of Microbiological Criteria for Foods and Food Ingredients*. National Academy Press, Washington, D.C.

Tompkin, R. B. 1992. Chapter 8. Corrective action procedures for deviations from the critical control point critical limits. In, Pierson, M. D. and D. A. Corlett, Jr. (eds.) *HACCP Principles and Applications*. Van Nostrand Reinhold, New York.

CHAPTER 10

RECORDKEEPING AND VERIFICATION PROCEDURES

by Bonnie J. Humm, K. E. Stevenson and John Y. Humber

PRINCIPLE 6:   **Establish Effective Recordkeeping Procedures that Document the HACCP System.**

INTRODUCTION

Records are written evidence through which an act is documented. Recordkeeping assures that this written evidence is available for review and is maintained for the required length of time.

Since part of the HACCP plan includes documentation relating to *all* critical control points (CCPs) identified in a food establishment operation, records are an integral part of a working HACCP system. All physical or chemical measurements of a CCP, any action on critical deviations and final disposition of any product, must be correctly documented and kept on file.

Records are the only reference available to trace the production history of a finished product. If questions arise concerning the product, a review of the records may be the only way to ascertain or even to prove that the product was prepared and handled in a safe manner in accordance with all the HACCP principles outlined in the company's HACCP plan.

The benefits derived from recordkeeping and review go beyond food safety. For example, it can be used as a tool or mechanism by which an operator may learn of an equipment or other malfunction and correct a potential problem that otherwise could lead to violation of a critical factor. Records of this type provide a dual function by providing a history of performance, as well as any actions taken to prevent a problem.

Record reviews must be conducted in-house by qualified staff members as well as by outside HACCP authorities, such as consultants, in order to assure strict compliance with the criteria set at the CCPs. Careful review of well-documented and maintained records is an invaluable tool in indicating possible problems and allowing corrective action to be taken before a public health problem occurs.

REASONS FOR KEEPING RECORDS

The reasons for keeping HACCP records relate to evidence of product safety with regard to the present procedures and processes, and the ease of product traceability and record review. Well maintained records provide irrefutable evidence that procedures and processes are being followed in strict accordance with HACCP requirements. Adherence to the specific critical limits (CLs) set at each CCP is the best assurance of product safety. Documenting the data of those measurements results in permanent records regarding the safety of products.

During regulatory compliance audits, company records may be the single most important source for data review. And, depending on the thoroughness of the records, the records may facilitate the work of the inspector in his attempt to ascertain the adequacy of processes and procedures used at the facility in question. More importantly, accurate records also provide plant personnel with this documentation of compliance.

Since HACCP records focus only on safety-related issues, problem areas can be identified quickly because these records provide an uncluttered view of product safety issues. It is best to keep all HACCP records separate from quality assurance documents so that only the product safety records will be viewed during HACCP audits. If a product safety problem occurs requiring a recall or market withdrawal, HACCP records assist in identifying lots of ingredients, packaging materials, and finished product which may be involved.

In order to assure product safety and to document processes and procedures, HACCP records must contain the following information:

- Title and date of the record
- Product identification (code, including time and date)
- Materials and equipment used
- Operations performed
- Critical criteria and limits
- Corrective action to be taken and by whom
- Operator identification
- Data (presented in an orderly format)
- The reviewer's initials and date of review

TYPES OF HACCP RECORDS

Critical Control Point Records

Records associated with establishing CCPs document the identification of specific hazards and the related preventive measures associated with each CCP. These hazards could be related to an ingredient, a packaging component or the process, and may be of a biological, chemical or physical nature.

A diagram or flow chart of the entire manufacturing process with each CCP identified would be a portion of these records. Since each CCP requires a risk assessment, proper documentation regarding the deliberations in determining the degree of risk associated with each hazard should be included in this category.

Once the CCPs have been assigned, a decision regarding the degree of control attainable at each CCP must be made. The criteria behind this decision should then be fully documented and made part of the HACCP records.

Records Associated with Establishing Critical Limits

In order to support the CLs established for each CCP, studies may have to be conducted and experimental data collected. The rational used to support the conclusions are important and should be included in this supporting data. In addition, any pertinent literature regarding the history of such criteria should also be included in this type of record. The precision and accuracy of all test methods used in the measurement of CLs must be well documented before making such tests part of the supporting documents for the HACCP program.

Records Associated with Monitoring CCPs

There are always normal and/or acceptable fluctuations in the data collected from most operations, and these fluctuations will be apparent on the records. It is crucial that the individual responsible for recording the CCP data knows the difference between normal fluctuations and an indication of loss of control at any CCP location. These guidelines must be clearly stated, and the CLs must be printed on each CCP record or data sheet for easy reference by the operator or attendant.

Examples of automated recording equipment include circular charts, recorders which document time and temperature, and electronic records of performance of metal detectors. Proper documentation of continuous monitoring systems will result in various charts, check lists and laboratory analysis sheets.

Discontinuous inspection, known as attribute sampling, is used primarily for chemical or physical testing, and the sampling rate should be based upon statistical data. This type of monitoring requires accurate documentation for each lot sampled.

Records Associated with Deviations

The failure to meet a required CL for a CCP is termed a deviation. The corrective action procedures for deviations must be documented in the HACCP plan. Each deviation requires a corrective action which must eliminate the actual or potential hazard and assure the safe disposition of the product involved. This requires a written record identifying the deviant lots. Non-compliant lots of product must be put on hold pending completion of the appropriate corrective actions, including a determination of appropriate disposition of the product. A Hold Summary could be the master form for these deviations, with supporting documentation kept in a separate file for a reasonable period after the expiration date of the product.

The final disposition and handling of all process or product deviations should be very detailed, including an accurate accounting of all units. This includes product destroyed, as well as product reworked or returned to stock.

Since HACCP deviations are related to product safety rather than quality, the HACCP records should be kept in a separate file apart from quality assurance or regulatory requirement records. This facilitates review of the HACCP records for compliance.

RECORD REVIEW

Records dealing with plant operations at CCPs must be reviewed on a daily basis; a designated, responsible individual must initial and date all records as they are reviewed.

This review must be followed up by a review of any deviations or irregularities. In addition, deviations from standard documentation procedures must be brought to the attention of the individuals responsible for filling out the reports, and these deviations must be corrected immediately.

Any anomalies must be investigated thoroughly for potential problems or trends, and, in this regard, record review becomes a preventive measure for assuring product safety. When this review reveals or identifies an inadequacy in the recordkeeping and normal monitoring procedures, existing parameters must be reviewed and updated. Because maintenance of the HACCP program results in a dynamic system, continual updating and improvements are necessary and warranted. For example, as the record review and verification process is applied to each component in the system, CCPs may be identified which were not addressed in the original program.

THE HACCP PLAN

The HACCP Plan is a written document which delineates the formal procedures to be followed in accordance with the seven HACCP principles. It may consist of a HACCP Manual or working document, appropriate HACCP test methods or SOPs, and a "Master File" containing all background documentation and HACCP records.

The HACCP Manual should include all the elements of the HACCP plan, such as:

- List of the HACCP team and assigned responsibilities.
- Description of the product and its intended use.
- Flow diagram for the entire manufacturing process indicating CCPs.
- Hazards associated with each CCP and preventive measures.
- Rational developed for determining hazard significance.
- Critical limit(s) for each CCP.
- Monitoring system, including sampling procedures and test methods.
- Corrective action plans for deviations from CLs.
- Recordkeeping procedures, including copies of forms and instructions.
- Procedures for verification of the HACCP system.

Plant personnel such as line workers or laboratory analysts should have copies of all SOPs or appropriate test methods for which they are responsible. This will enable them to properly execute their individual HACCP assignments.

Any revisions of the HACCP plan must be immediately reflected in the HACCP Manual, thus, document control is important. All charts should have issue numbers so CLs and instructions are kept current. Outdated sections and documentation forms should be immediately discarded to avoid confusion. A periodic review of departmental HACCP forms and procedures may be necessary to assure compliance with the HACCP plan.

When revisions are made and sent to their respective departments, it is advisable to have routing slips attached so that the individuals responsible for the implementation of those revisions are properly notified. There is nothing worse than having outdated HACCP documents still on file and in use at a facility purporting to be operating under a comprehensive HACCP system.

Investigative reviews should be conducted by staff personnel prior to audits in order to identify weaknesses in the documentation or recordkeeping system. Having a well organized system for documentation will show that a company is in control of the operation, in general, and is in control of product safety issues specifically. It is important that all in-house record reviews be well documented with all deficiencies noted and corrective action(s) clearly outlined. When problems continue to occur in a certain area, there must be a written record of the cause(s) and the solution(s).

RETENTION OF RECORDS

Where regulatory requirements for retention of records exist, they vary depending upon the product, regulatory agency and locale. HACCP records should be held for at least a year, while any records required by law to be kept longer than one year should be kept the legally mandated period. The shelf-life of products also needs to be taken into account in establishing retention guidelines. FDA's low acid canned foods regulations specify that copies of all required thermal processing records, records of pH measurements, process deviations and other "critical factors" shall be retained at the processing facility for one year, and for two additional years in an accessible location.

REGULATORY ACCESS

The question of regulatory access to HACCP records will be specifically addressed in pending regulatory activities. The types of records utilized in the total HACCP system include pertinent records on ingredients and packaging materials, processes and controls, packaging requirements, storage and distribution. Records that deal with the management or function of the system itself and proprietary information would not normally be made available to the regulatory agencies. Records that clearly relate to product safety are already identified in the HACCP program and may be subject to the scrutiny of regulatory authorities. Having these records well organized makes data retrieval an easy task for both internal and external audits.

RECORDKEEPING PROCEDURES

Personnel responsible for documenting HACCP records should never pre-record data in anticipation of the actual data, or postpone making entries and rely on their memories. These records may be the company's only proof that a CCP was controlled or that appropriate corrective action was taken to assure the safety of the product. Thus, they must be kept in a timely and accurate fasion.

Any modifications to the existing data should never be erased, but lined out and corrected with the responsible individual's initials alongside the change.

To be used effectively, HACCP records should be on standardized forms for the company and must be reviewed regularly by a responsible individual for completeness. A thorough review must ensure that all critical factors have been satisfied and are accurately documented.

Management, supervisors and inspectors are all responsible for the safety of our food products and therefore have a primary role in assuring that all HACCP records are accurate and complete, and that these records reflect the actual operating conditions. Assuring compliance with existing and newly promulgated regulations requires each processor to keep updated on the regulations governing product safety.

PRINCIPLE 7: **Establish Procedures for Verification that the HACCP System is Working Correctly.**

INTRODUCTION

Verification is a process whereby compliance with the HACCP Plan is evaluated. It involves actual observation of prescribed procedures and a thorough review of records via internal and external audits. The verification process is designed to review the HACCP Plan, to establish whether the CCPs and CLs are being adequately controlled and monitored, and to determine if the procedures for product deviations and recordkeeping are being followed correctly. Whether the verification review comes from within the establishment or from an outside authority, the basic objectives are the same--to evaluate (a) the development of, and (b) the day-to-day compliance with the HACCP Plan.

HACCP SYSTEM EVALUATION

The purpose of the evaluation of a HACCP system is to ensure that all hazards have been identified, and that every hazard is being controlled to the degree necessary (through appropriate CLs) and thereby provide the consumer with a safe product.

Events which Trigger Evaluation

An evaluation of the HACCP system should be implemented when one or more of the following events occur:

- New information concerning the safety of the product becomes available.
- The food product or product category is linked to a foodborne disease outbreak.
- The production system has been modified.

- A specified length of time has passed since the last HACCP evaluation.
- A change occurs in the formulation, production, distribution or consumer use of the product.

Identification of Potential Deficiencies

Potential deficiencies in an existing HACCP system can be identified by reviewing the resources listed below. In addition to confirming that an evaluation of the HACCP system is warranted, a review of these resources can provide the HACCP team with information on hazards and appropriate limits for CCPs. If necessary, CLs can be revised based on the new information.

- HACCP records, including flow diagrams, deviations from CLs, and instrument calibration records.
- Product history, including recall and market withdrawal information.
- Centers for Disease Control *Morbidity and Mortality Weekly Report* summaries of foodborne outbreaks.
- Process authority recommendations.
- Predictive model data used to assess product/process safety.
- Scientific literature articles.
- Regulatory agency alerts.
- Test results from sample monitoring.

Phases of the Evaluation

The first phase of the evaluation verifies that all critical areas related to product safety have been identified, and that each potential hazard has a designated CCP. Also, CCPs should be reviewed to determine if each previously identified CCP is still deemed necessary. As food production lines change, new CCPs may need to be added while some old CCPs are not needed and thus become obsolete.

The second phase of a HACCP evaluation should concentrate on the CLs previously established for each CCP. It is necessary to confirm that the current CLs are appropriate and reasonable, to ensure product safety. This phase should also confirm that the monitoring method used for each CCP and the frequency of monitoring of each CCP are effective in detecting deviations from CLs, if deviations exist.

The final part of the evaluation should establish that corrective action procedures are in place and are effective in eliminating any deficiencies in the HACCP system. If additional CCPs are needed, they should be added. If deviations from CLs exist, they should be corrected to the degree necessary to ensure that they are consistently within acceptable tolerances. In some cases, it may be necessary to either relax or tighten the CLs of an existing CCP.

INTERNAL VERIFICATION

An important key to the successful evaluation and revision of a HACCP system is the interaction of the HACCP team. The HACCP team often includes representatives from a number of different departments. Working together, team members with appropriate knowledge and expertise can help to ensure that the consumer will consistently receive a safe food product.

Review of Records

All CCP records should be reviewed daily by management to assure compliance with the existing parameters or CLs, and to determine if there were any problems or anomalies which need to be investigated and/or corrected. Proper follow-up on product holds is also recommended. Other records regarding product safety issues, such as those dealing with sanitation records and GMP compliance should also be reviewed on a regular schedule to assure compliance with all necessary practices.

All in-house verification procedures should be recorded by the individuals reviewing these records by either (a) initialing and dating the original records, or (b) filling out a "HACCP Verification Report" designed by the individual institution for documenting this step in the implementation of the HACCP program.

The individuals responsible for the review of HACCP documentation must be knowledgeable of the HACCP principles and the role they play in its verification process. When corrections are needed to improve the present system these individuals should have the authority to suggest or make corrections. The individuals responsible for verification could be from Quality Assurance, Production or the Maintenance Department, depending on the specific nature of the CCP records.

HACCP Plan Revisions

Since any additions to the product line or equipment should be reviewed before being integrated into the original HACCP plan, the following changes should require committee review:

1. Introduction of a new product or a change in formulation to any existing product.
2. Any change in a processing parameter.
3. Installation of any line equipment or modification.
4. Changes in packaging or final product handling.
5. Change in the intended handling by the consumer (frozen to refrigerated).
6. Label changes regarding preparation of the product by the consumer.
7. New hazard awareness regarding potential pathogens or environmental contaminants.
8. Use of product by populations "at risk".

In essence, many changes can affect the safety of a product. It is therefore essential to obtain prior review and approval for any and all changes proposed in the product or process.

Periodic Verification Procedures

Spot checks or sample analyses are also established verification procedures which may require specific documentation from the processor regarding lot or code numbers. These checks may serve to validate an existing ingredient warranty or guarantee kept on file, and also serve to substantiate the assignment of certain CCPs in the HACCP system. In addition, the results of such analyses provide evidence of the adequacy of present procedures used in-house for these critical factors.

Since proper calibration of existing equipment is critical to the accuracy and precision of any analyses, review of calibration records may alert plant personnel of an existing or potential problem so corrective action can be taken. In-house verification procedures require comprehensive record reviews on a routine basis in order to assure full compliance with the HACCP program.

EXTERNAL VERIFICATION

In some instances, regulatory agencies may conduct audits of HACCP plans to verify that the HACCP system is operating effectively to control product safety hazards. Once the HACCP system is in place, a regulatory agency audit may include, but is not limited to, the following activities:

- Review of the written HACCP plan.
- Review of records concerning CCPs.
- Review of deviations and dispositions.
- Visual observations of operations.
- Random sample collection and analysis.
- Written records of verification audits and corrective actions taken.

Normally, a written record of the findings of these audits will be given to the company with requests that corrective action(s) be taken regarding any problems found. These regulatory verification audit reports should be responded to in writing and maintained as part of the HACCP documentation program.

Due to the dynamic nature of the HACCP plan, this includes daily checks by staff to assure compliance, routine reviews leading to modifications which make the existing program even more effective, and occasional internal and external evaluations of the entire HACCP Plan to verify the system is working effectively.

REFERENCES

CFDRA. 1987. *Guidelines to the establishment of Hazard Analysis Critical Control Point (HACCP)*. Campden Food and Drink Research Association. Gloucestershire, England.

Ellinger, R.H. 1990. *Total quality systems handbook - HACCP*. American Butter Institute/National Cheese Institute. Washington, DC.

Humber, J. 1992. Control points and critical control points. In, Pierson, M. D. and D. A. Corlett, Jr. (eds.), *HACCP - Principles and Applications*. Van Nostrand Reinhold. New York.

IAMFES. 1991. *Procedures to implement the Hazard Analysis Critical Control Point System*. International Association of Milk, Food and Environmental Sanitarians, Inc. Ames, IA.

ICMSF. 1988. *Microorganisms in foods 4: Application of the hazard analysis critical control point (HACCP) system to ensure microbiological safety and quality*. The International Commission on Microbiological Specifications for Foods. Blackwell Scientific Publications. Oxford, England.

Moberg, L. 1990. *HACCP verification*. Presented at the American Association of Cereal Chemists HACCP short course. Chicago.

NACMCF. 1992. Hazard Analysis and Critical Control Point System. *Intl. J. Food Microbiol.* 16:1.

Prince, G. 1992. Verification of the HACCP program. In, Pierson, M. D. and D. A. Corlett, Jr. (eds.), *HACCP - Practices and Applications*. Van Nostrand Reinhold, New York.

Stevenson, K. E. and B. J. Humm. 1992. Effective recordkeeping system for documenting the HACCP plan. In, Pierson, M. D. and D. A. Corlett, Jr. (eds.), *HACCP - Principles and Applications*. Van Nostrand Reinhold, New York.

CHAPTER 11

WORKSHOP FLOW DIAGRAMS and FORMS

by K. E. Stevenson and Allen M. Katsuyama

INTRODUCTION

The materials provided in this chapter were designed for use in workshops on the application of HACCP Principles. This includes procedures for conducting a microbiological risk assessment; production flow diagrams for five model products; and forms for use in the application of HACCP Principles.

MICROBIOLOGICAL RISK ASSESSMENT

The Pillsbury Company developed a two-step "microbiological risk assessment" to assist in the hazard analysis for microbiological hazards. In 1989, the National Advisory Committee on Microbiological Criteria for Foods (NACMCF) expanded the original microbiological risk assessment and included this expanded version as part of the description of how to apply Principle 1. In the latest version of the HACCP Principles, this microbiological risk assessment was taken out of Principle 1 and placed in the Appendix as reference material (See Chapter 2, Appendix 2-B; and the discussion in Chapter 8). For the reasons outlined in Chapter 8, we have **not** used the exact format outlined below in recent workshops. We present it here because the process of working through this procedure provides perspective on the type of reasoning applied during the assessment.

The microbiological risk assessment procedure which was included in the 1989 NACMCF document consisted of a two-step process, (a) ranking an ingredient or food according to six hazard characteristics, and (b) assignment of a risk category based upon the results of the initial rankings. This logic may be useful in determining if a hazard is of sufficient importance to warrant consideration for control with the HACCP plan. Ingredients and finished products can be subjected to this procedure. If applied to ingredients, the food plant should be considered the "consumer."

Microbiological Hazard Analysis

Rank the food according to the general hazard characteristics A through F, using a plus (+) to indicate a potential hazard. Use a zero (0) to indicate the hazard characteristic does not apply. The number of pluses will be used later to determine the risk category.

The general hazard characteristics developed by the NACMCF are listed in Table 11-1. An alternative list is presented in Table 11-2 for use with ingredients.

TABLE 11-1. MICROBIOLOGICAL HAZARD CHARACTERISTICS

Hazard A: A special class that applies to non-sterile products designated and intended for consumption by at-risk populations, e.g., infants, the aged, the infirm, or immunocompromised individuals.

Hazard B: The product contains "sensitive ingredients" in terms of microbiological hazards.

Hazard C: The process does not contain a controlled processing step that effectively destroys harmful microorganisms.

Hazard D: The product is subject to recontamination after processing before packaging.

Hazard E: There is substantial potential for abusive handling in distribution or in consumer handling that could render the product harmful when consumed.

Hazard F: There is no terminal heat process after packaging or when cooked in the home.

TABLE 11-2. ALTERNATIVE MICROBIOLOGICAL HAZARD CHARACTERISTICS FOR USE WITH INGREDIENTS

Hazard A: A special class that applies to non-sterile ingredients for use in non-sterile products designated and intended for consumption by at-risk populations, e.g., infants, the aged, the infirm, or immunocompromised individuals.

Hazard B: The ingredient contains "sensitive materials" in terms of microbiological hazards.

Hazard C: The supplier's process for the ingredient does not contain a controlled processing step that effectively destroys harmful microorganisms.

Hazard D: The ingredient is subject to recontamination (at the supplier) after processing before packaging. (Ingredients thermally processed in their containers, e.g., canned ingredients, are rated "0".)

Hazard E: Prior to use in your facility, there is substantial potential for abuse in distribution or in handling the ingredient that eventually could render your product harmful.

Hazard F: There is no terminal heat process after packaging at the supplier or when used at your facility.

Assignment of Risk Categories

If the product or ingredient is "+" for Hazard A, it is automatically assigned to Risk Category VI. If the product or ingredient is "0" for Hazard A, the risk category is determined simply by the number of pluses assigned for Hazards B through F. Explanations of the microbiological risk categories are presented in Table 11-3.

TABLE 11-3. MICROBIOLOGICAL RISK CATEGORIES

Category VI:	A special category that applies to nonsterile products designated and intended for consumption by at-risk populations, e.g. infants, the aged, the infirm, or immunocompromised individuals. All six hazards must be considered.
Category V:	Food products subject to all five general hazard characteristics. Hazard Characteristics B, C, D, E, and F.
Category IV:	Food products subject to four of the general hazard characteristics.
Category III:	Food products subject to three of the general hazard characteristics.
Category II:	Food products subject to two of the general hazard characteristics.
Category I:	Food products subject to one of the general hazard characteristics.
Category 0:	Hazard Class--No hazard.

Use of Microbiological Risk Assessment Data

Although there is no specified procedure which directly links the microbiological risk categories with identification of critical control points (CCPs), this procedure provides valuable information to help assess the significance of hazards before identifying CCPs. First, each "+" for a hazard characteristic provides a warning that the potential risk associated needs to be evaluated with respect to that specific type of hazard. Second, if a product/process is assigned Risk Category I or II, it apparently represents a relatively low risk. While the potential health hazard needs to be evaluated, it is unlikely that the hazard is of great enough significance that it must be addressed within the HACCP plan. Thus, a CCP will should not be required to adequately control this risk. Hazards of such low risk are typically controlled within the context of prerequisite programs such as basic hygienic practices. However, if the product or process is assigned Risk Category V, it may represent a relatively high risk. In this instance, a CCP may be necessary to eliminate or reduce the potential microbiological hazard associated with the product/process.

One note of caution, no scheme has yet been devised which allows for a **quantitative** assessment of risk from microbiological hazards. Thus all methods of assessment, including the one described above will be qualitative at best and they must be viewed as rough approximations of the importance of a particular hazard. Common sense is still the most important tool in evaluating results of the hazard/risk assessment exercise.

Presentation of Microbiological Risk Assessment Data

Table 11-4 provides an example of the format for a chart for use in assessment of a food or ingredient by hazard characteristic and risk category (as presented in the NACMCF HACCP document).

TABLE 11-4. PRODUCT RISK CHARACTERISTICS BASED UPON MICROBIOLOGICAL HAZARD CHARACTERISTICS

Food Ingredient or Product	Hazard Characteristics (A, B, C, D, E, F)	Risk Category
T	A+ (Special Category)	VI
U	Five +'s (B through F)	V
V	Four +'s (B through F)	IV
W	Three +'s (B through F)	III
X	Two +'s (B through F	II
Y	One + (B through F)	I
Z	No +'s	O

Table 11-5 presents information for use in a workshop on conducting a microbiological risk assessment for a model product, Rocky Road Ice Cream, and some typical ingredients for this type of product. The hazard characteristics and risk categories have been completed for the product and some of the ingredients. The variability (O/+) associated with some of the hazard categories represents the possibility of the use of different handling procedures by the supplier. Knowledge of the conditions and procedures used by the supplier is important in determining the actual hazard characteristics for each of the ingredients.

(Note: The flow diagram for this product is presented in Figure 11-1.)

TABLE 11-5. MICROBIOLOGICAL RISK ASSESSMENT FOR ROCKY ROAD ICE CREAM

Item	Special Category	Sensitive Ingredient	Microbes Not Destroyed	Recontam-ination Potential	Abuse Potential	No Terminal Heat Process	Hazard Category
	A	B	C	D	E	F	
PRODUCT							
Ice Cream	0	+	0	+	0	+	III
INGREDIENTS							
Fresh Cream							
Whey	0	+	+	+	+	0	IV
Milk							
Sugar	0	0	0	0/+	0	0	0/I
Choc. Liquor	0	+	+	+	0	0	III
Cocoa Powder							
Marshmallows							
Nuts							
Gums							
Artificial Color	0	0	+	0/+	0	0	II/I

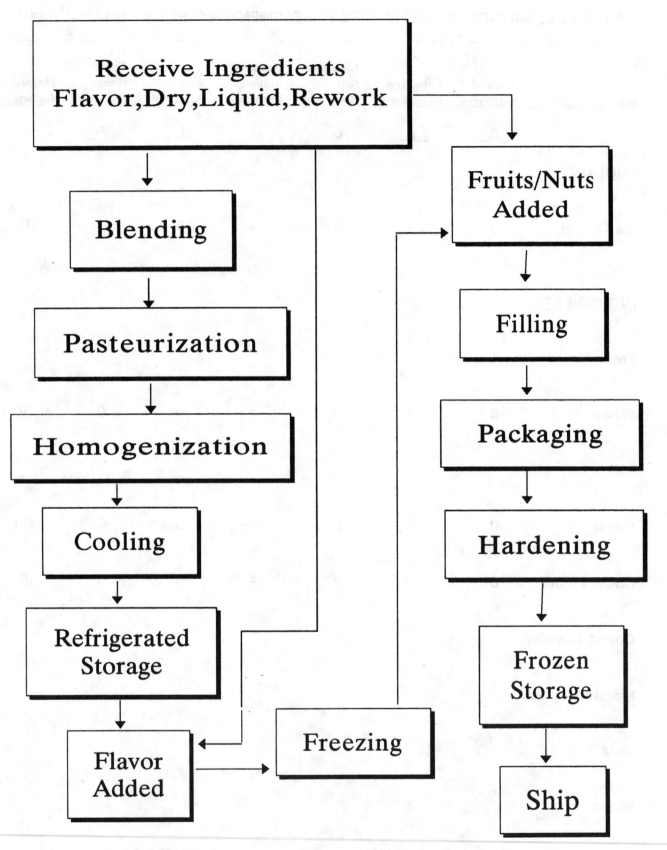

Figure 11-1. FLOW DIAGRAM FOR ROCKY ROAD ICE CREAM

WORKSHOP ON APPLICATION OF HACCP PRINCIPLES

The remainder of this Chapter presents additional materials for use in workshops on the application of HACCP Principles to food processing operations.

Table 11-6 provides examples of potential physical hazards and related CCPs associated with a model dry goods product--buttermilk pancake mix. (Note: These CCPs identified in this model are specific for this particular operation and may not reflect an assessment of significant hazards in all cases. No attempt has been made to include potential microbiological and chemical hazards and related CCPs in this workshop example.) The flow diagram for production of the buttermilk pancake mix is presented in Figure 11-2.

FORMS

Three forms are provided (Forms A, B, and C) which are useful for developing information during workshops on the application of HACCP Principles to various food processing operations:

FORM A: HAZARD ANALYSIS AND CCPs

Provides a format for documentation related to conducting a hazard analysis and identifying CCPs.

FORM B: CRITICAL LIMITS, MONITORING, AND CORRECTIVE ACTIONS

Once the CCPs have been identified, this form can be used to document the critical limits (CLs) for each CCP, as well as, some of the important elements associated with monitoring, and examples of corrective actions.

FORM C: RECORDKEEPING AND VERIFICATION

This form can be used to list the specific records which should be kept in association with each CCP, and verification procedures which are specific to the individual CCPs.

HACCP Master Sheet

The information supplied on Forms A, B and C is commonly used to prepare a HACCP "Master Sheet." The HACCP Master Sheet provides an overview of the HACCP Plan, and is helpful in disseminating a summary of the HACCP Plan to individuals who are not interested in detailed information. (This type of summary will probably be read by more people than any other document related to the HACCP Plan.) Completed HACCP Master Sheets are provided for three model products. See examples 11-1, 11-2 and 11-3.

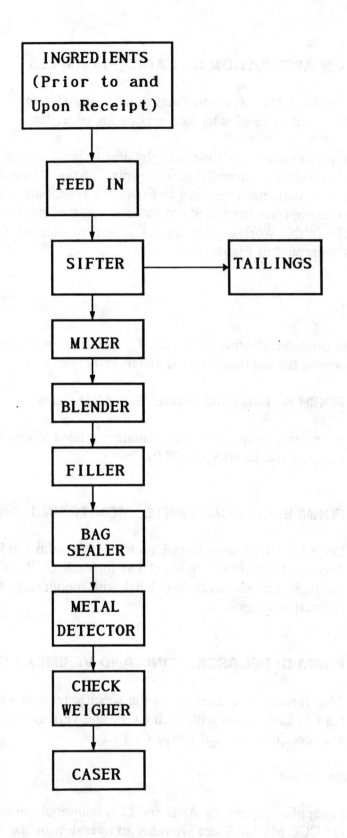

Figure 11-2. FLOW DIAGRAM FOR BUTTERMILK PANCAKE MIX

TABLE 11-6. HAZARDS, PREVENTIVE MEASURES AND CRITICAL CONTROL POINTS FOR BUTTERMILK PANCAKE MIX.

Process Step	Potential Hazard	Preventive Measure	CCP?/Description
01 Feed-In	Foreign material	Inspect each bag.	No (CP)/Visual inspection of bags
02 Sifter Screen	Metal	Inspect screen for damage/holes.	No (CP)/Visual inspection of screen for metal/ holes.
03 Tailings	Foreign material	Inspect tailings. Label, and retain material for evaluation.	Yes/Visual inspection of tailings for type of foreign material.
04 Magnets	Tramp metal	Inspect magnets. Retain and label metal for evaluation.	No (Metal detector is CCP.)
05 Scalping Screen	Foreign material	Inspect screen for damage/holes.	No (CP)/Visual inspection of screen for damage and presence of foreign material.
06 Filler Tooling	Metal fragments	Inspect filler for scoring or chipping of metal surfaces.	No (Metal detector is CCP.)
07 Metal Detector	Metal	Metal detector operating properly. Metal from "kick-outs" retained and labeled.	Yes/Detection of metal fragments using metal detector.

FORM A

HAZARD ANALYSIS AND CCPs

Process Step	Potential Hazard	Preventive Measure	CCP?/Description

FORM B

CRITICAL LIMITS, MONITORING, AND CORRECTIVE ACTIONS

Process Step/CCP	Critical Limits	Monitoring Procedure				Corrective Action
		What	How	Frequency	Who	

FORM C

RECORDKEEPING AND VERIFICATION

Process Step/CCP	Records	Verification Procedures

HACCP MASTER SHEET

Critical Control Point (CCP)	Hazard	Critical Limits of the Preventive Measures	Monitoring				Corrective Action	Records	Verification
			What	How	Frequency	Who			

EXAMPLE 11-1

FOR ILLUSTRATIVE PURPOSES ONLY

Process Flow Description
for Battered and Breaded Chicken Pieces

The chicken is fully cooked by deep fat frying. A PET tray is used to pack 6, 8, or 12 pieces using a heat sealed plastic film lid. There is no atmosphere modification. PET trays are packed in cartons for distribution. Product is intended for general consumption. Shelf life is 15 days with "used by" date and "keep refrigerated" on label.

Refrigerated, raw, pre-sized, skinless chicken pieces are received via refrigerated truck. The chicken pieces are transferred to refrigerated storage for holding prior to processing. The refrigeration units are equipped with high temperature alarms.

Dry ingredients (batter, bread crumbs and spices) are received and placed in dry goods storage. Frying oil is also received and stored in stainless steel tanks.

Individual chicken pieces pass through an automated checkweigher and are transferred to the batter coating tank. Pieces which are too large or too small are rejected. Dry batter mix is blended with water and held in a tank for addition to the batter coating machine. The batter in the holding tank is chilled to 45°F. The tank is equipped with a high temperature alarm. The individual pieces are spray coated with batter, inverted, and spray coated on the other side.

The breading is applied in a manner similar to the batter. The battered pieces are breaded, inverted, and the opposite side of the piece is breaded.

After breading, the pieces pass through a continuous fryer. The fryer temperature is set at 450°F. The frying time is determined by the belt speed in the fryer. The chicken pieces are completely cooked to an internal temperature of 165°F. Testing has been conducted by a process authority to verify that the time-temperature relationship used to operate the fryer is adequate to achieve an internal temperature of 165°F as long as piece weight does not exceed 4.5 oz.

The cooked pieces pass through a cooling tunnel and exit at an internal temperature of 40°F or less. The pieces are inspected for breading uniformity and/or damage. Cooling studies were conducted to verify that operating the cooler at 0°F with a belt speed of $\leq$ 4 feet/minute will result in an internal temperature of $\leq$ 40°F within 4 hours.

The pieces are individually packaged in 6, 8, or 12 packs, coded with a day code, and passed through a metal detector. Packages are cased, palletized and transferred to refrigerated storage. The refrigerator is set for a maximum temperature of 40°F and is equipped with a high temperature alarm. Product is shipped via refrigerated truck to a distribution center.

EXAMPLE 11-1

FOR ILLUSTRATIVE PURPOSES ONLY

BATTERED & BREADED CHICKEN PIECES

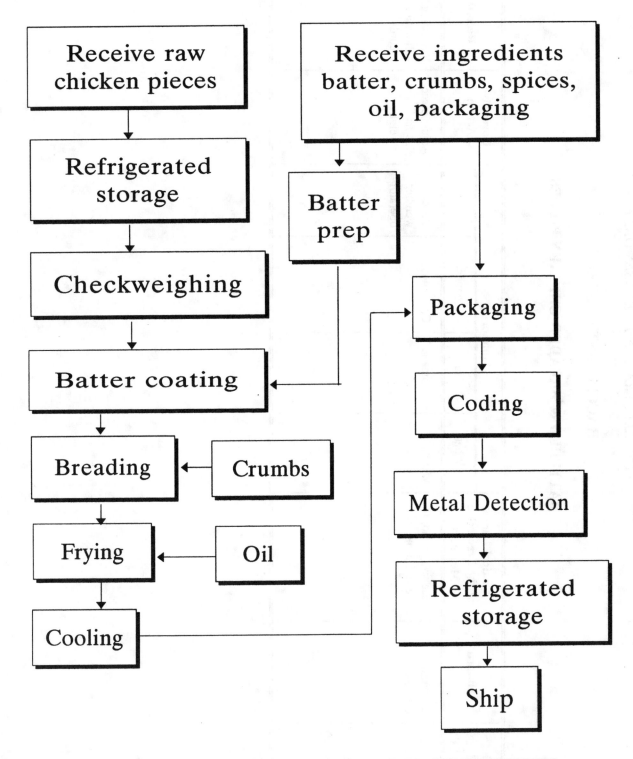

Figure 11-3. FLOW DIAGRAM FOR BATTERED AND BREADED CHICKEN PIECES

EXAMPLE: 11-1

FOR ILLUSTRATIVE PURPOSES ONLY

HACCP MASTER SHEET

BATTERED & BREADED CHICKEN PIECES

Critical Control Point (CCP)	Hazard	Critical Limits of the Preventive Measures	Monitoring				Corrective Action	Records	Verification
			What	How	Frequency	Who			
Check-weigher CCP1 (B)	Bacterial pathogens	Piece weight not to exceed 4.5 oz.	Weight	Automatic check-weigher	Continuous	Check-weigher operator	Adjust piece size and reweigh	Reject log, calibration records	Operator will run check sample at startup and once each hour. Q.A. will calibrate checkweigher monthly.

FOR ILLUSTRATIVE PURPOSES ONLY

HACCP MASTER SHEET

BATTERED & BREADED CHICKEN PIECES

Critical Control Point (CCP)	Hazard	Critical Limits of the Preventive Measures	Monitoring				Corrective Action	Records	Verification
			What	How	Frequency	Who			
Frying CCP2(B)	Bacterial pathogens	Oil temperature ≥ 450°F	Temperature	Temperature recorder, low temperature alarm	Continual recording	Fryer operator checks and initials recorder chart	Hold product if low temperature. Notify Q.A. Evaluate effect of low temperature, adjust fryer if necessary.	Fryer operator log, recorder chart, record of internal temp. of chicken, QA calibration records.	Q.A. will calibrate recorder and alarm each month. Daily record review and initial. Check oil temp. each shift and compare with recorder. Measure internal temp. of chicken.
		Chain speed ≤ 4 feet per minute	Time	Belt speed tachometer on drive shaft	Continual recording of speed	Fryer operator checks and initials recorder chart	Hold product if belt speed > 4 feet per minute and evaluate.	Fryer operator log, recorder chart, QA calibration records, verification records	Q.A. will calibrate tachometer every 2 weeks. Daily record review and initial. Operator measures belt speed once per shift according to SOP.

11-17

EXAMPLE 11-1

FOR ILLUSTRATIVE PURPOSES ONLY

HACCP MASTER SHEET

BATTERED & BREADED CHICKEN PIECES

Critical Control Point (CCP)	Hazard	Critical Limits of the Preventive Measures	Monitoring				Corrective Action	Records	Verification
			What	How	Frequency	Who			
Cooling CCP3(B)	Growth of bacterial pathogens	Cooler temperature 0°F*	Temperature of cooler	Temperature recorder, high temperature alarm, thermometer	Continual	Cooler operator checks recorders at beginning and end of shift; checks and records temperature on dial thermometer each hour	Reset cooler temperature and notify maintenance if necessary. Isolate product and immediately re-cool to specification or hold/destroy.	Cooler operator's log, verification records (calibration, etc.)	QA will conduct daily record review and initial; calibrate recorder and alarm; verify internal temp. of chicken exiting cooler daily.
		Belt speed ≤ 4 ft. per min.*	Minimum cooling time	Belt speed tachometer (see belt speed monitoring procedure)	Continual recording	Cooler operator checks chart at beginning of shift and initials at end	If belt speed > 4 ft. per min isolate, hold refrigerated. Evaluate to determine disposition. Fix cooler	Cooler operator's log, verification records (calibration, etc.)	Q.A. will calibrate tachometer each month, confirm belt speed at start-up

* Note that product internal temperature during cooling are preventative measures for pathogens in this product. The model presented allows for monitoring of those critical limits which directly influence internal temperature. An alternative approach would be to directly monitor internal temperature.

FOR ILLUSTRATIVE PURPOSES ONLY

HACCP MASTER SHEET

BATTERED & BREADED CHICKEN PIECES

Critical Control Point (CCP)	Hazard	Critical Limits of the Preventive Measures	Monitoring				Corrective Action	Records	Verification
			What	How	Frequency	Who			
Metal detector CCP4(P)	Metal	Operable metal detector	Ferrous and non-ferrous metal	Automatic screening	Continual	Labeling operator checks to ensure detector is on	Line operator evaluates rejects hourly. Excessive reject rate forces line shutdown. Q.C. prepares deviation report, determines source of metal, and corrects it. Contact maintenance if necessary - fix detector. If detector is not on, or fails sensitivity check, all product since last acceptable check is held and rechecked for metal. Operator prepares deviation report, notify Q.C./Q.A.	Operator log indicating time of metal detector sensitivity checks and results. Deviation reports indicating lot and time when metal was detected or detector failed, results of evaluation, and disposition of product. Metal detector maintenance and calibration records.	Run test material with metal of appropriate size to check sensitivity hourly (label operator). Supervisor reviews operator logs daily, deviation reports weekly.

EXAMPLE: 11-1

FOR ILLUSTRATIVE PURPOSES ONLY

HACCP MASTER SHEET

BATTERED & BREADED CHICKEN PIECES

Critical Control Point (CCP)	Hazard	Critical Limits of the Preventive Measures	Monitoring				Corrective Action	Records	Verification
			What	How	Frequency	Who			
Refrigerated storage (cooler) CCP5(B)	Pathogen growth	$\leq 40°F$	Cooler temperature	Temperature recorder, high temperature alarm, thermometer	Continual recording, intermintant check of thermometer	Shipping clerk	Isolate, hold refrigerated, evaluate to determine disposition	Cooler log, recorder chart, verification records	Q.A. will calibrate alarm; review log daily; internal temperature check daily; audit.

EXAMPLE 11-2

FOR ILLUSTRATIVE PURPOSES ONLY

Process Flow Description
for Raw Frozen Beef Patties With Textured Vegetable Protein

The product is packed in plastic-lined boxes for distribution to food service institutions. Each box is labelled with "use by" date and the instruction to "keep refrigerated."

Frozen 50 lbs. blocks of lean and regular beef trimmings are received from domestic sources, inspected, and placed into refrigerated storage (40ºF) for tempering. The refrigeration units are equipped with high temperature alarms and recorder charts. The meat has been previously graded and sorted according to processors' specifications.

The tempered meat is inspected and the regular and lean meat are ground separately. The regular ground meat is mixed back into the lean ground meat and blended according to customer specifications. Rehydrated TVP is blended into the mix.

The ground meat is transferred to the patty forming machine. Individual patties are conveyed into a blast freezer where they are frozen to an internal patty temperature of less than 20°F at the freezer exit.

Frozen patties are packaged (48/box), labeled, and palletized. The pallets are transferred to frozen storage. The storage freezer is set at -10°F and has a high temperature alarm. Pallets are shipped to a distributor.

EXAMPLE 11-2

FOR ILLUSTRATIVE PURPOSES ONLY

RAW FROZEN BEEF PATTIES WITH TVP

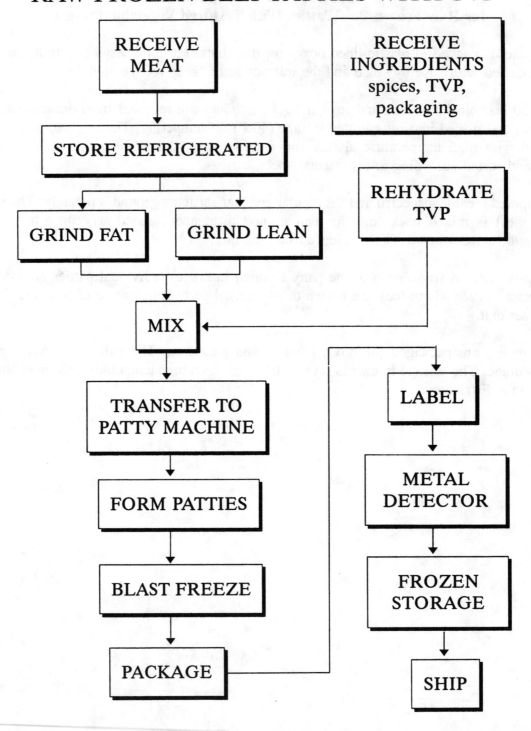

Figure 11-4. FLOW DIAGRAM FOR RAW FROZEN BEEEF PATTIES WITH TEXTURED VEGETABLE PROTEIN

EXAMPLE: 11-2

FOR ILLUSTRATIVE PURPOSES ONLY

HACCP MASTER SHEET

RAW FROZEN BEEF PATTIES WITH TEXTURED VEGETABLE PROTEIN

Critical Control Point (CCP)	Hazard	Critical Limits of the Preventive Measures	Monitoring				Corrective Action	Records	Verification
			What	How	Frequency	Who			
Receiving meat CCP1(B)	Bacterial pathogens	Internal temperature ≤ 40°F Acceptable organoleptic condition	Temperature, odor, color	Thermometer, visual, organoleptic inspection	4 boxes selected at random from each lot	Receiving clerk	Hold refrigerated and evaluate or reject	Receiving log, calibration log	Q.A. will calibrate thermometer weekly and, review records (initial) daily.
Label CCP2(B)	Bacterial pathogens	Label with proper cooking instructions	Label present	Label scanner	Continual	Label operator	Reject and relabel	Labeling operator log to indicate proper labels loaded and scanner functioning, verification records	Q.A. will verify scanner function w/ unlabeled packages every 2 hours, visual check of labeled packages hourly

11-23

EXAMPLE: 11-2
FOR ILLUSTRATIVE PURPOSES ONLY
HACCP MASTER SHEET
RAW FROZEN BEEF PATTIES WITH TEXTURED VEGETABLE PROTEIN

Critical Control Point (CCP)	Hazard	Critical Limits of the Preventive Measures	Monitoring				Corrective Action	Records	Verification
			What	How	Frequency	Who			
Metal detector CCP3(P)	Metal	Operable metal detector	Ferrous and non-ferrous metal	Automatic screening	Continual	Labeling operator checks to ensure detector is on	Line operator evaluates rejects hourly. Excessive reject rate forces line shutdown. Q.C. prepares deviation report, determines source of metal, and corrects it. Contact maintenance if necessary - fix detector. If detector is not on, or fails sensitivity check, all product since last acceptable check is held and rechecked for metal. Operator prepares deviation report, notify Q.C./Q.A.	Operator log indicating time of metal detector sensitivity checks and results. Deviation reports indicating lot and time when metal was detected or detector failed, results of evaluation, and disposition of product. Metal detector maintenance and calibration records.	Run test material with metal of appropriate size to check sensitivity hourly (label operator). Supervisor reviews operator logs daily, deviation reports weekly.

EXAMPLE: 11-2

FOR ILLUSTRATIVE PURPOSE ONLY

HACCP MASTER SHEET

RAW FROZEN BEEF PATTIES WITH TEXTURED VEGETABLE PROTEIN

Critical Control Point (CCP)	Hazard	Critical Limits of the Preventive Measures	Monitoring				Corrective Action	Records	Verification
			What	How	Frequency	Who			
Frozen storage CCP4(B)	Bacterial pathogens	Internal temperature ≤ 40°F*	Temperature	High temperature alarm freezer	Continual	Temperature monitor	Hold frozen, evaluate hazard	Freezer log, calibration check log	Q.A. will calibrate alarm, check monthly

* As an operational limit, product should be kept frozen. Temperature control monitored using freezer temperature.

EXAMPLE 11-3

FOR ILLUSTRATIVE PURPOSES ONLY

**Process Flow Description
for Breaded Fish Sticks**

The fish sticks are not fully cooked. These are packed in a PET tray with a heat sealed plastic film lid. Th[is] is no atmosphere modification. Each package is labelled with a "use by" date, cooking instructions, and[an] instruction to "keep frozen." The product is intended for the general public.

Imported frozen minced fish (either pollack or haddock) is received in frozen blocks via freezer truck. [The] blocks are transferred to frozen storage, the freezer is set at -10°F and monitored by a recording chart [and] alarm system.

Dry ingredients (batter, breading) and packaging materials are delivered to the plant by truck. Dry goods [are] placed in dry, cold storage.

To be processed, the fish blocks are removed from the freezer, one pallet at a time. Cases are opened [and] blocks unwrapped. Blocks are cut into pre-formed fish sticks with a saw. As sticks proceed on a conve[yor] belt, they are culled for uniformity and then battered and breaded, twice each. Batter is kept chilled to 45[°F] to prevent potential growth of pathogenic microorganisms. Batter temperature is monitored hourly by Qua[lity] Control.

From the last breading application, the portions pass through a fryer containing soy bean oil for less than [a] minute at 400°F. This fryer sets the batter/breading but does not cook the fish.

The fish sticks exit the fryer and enter a nitrogen tunnel for individual quick freezing. The nitrogen tun[nel] freezer is set at temperatures equivalent to -120°F; the exposure time is 6-10 minutes.

As the fish sticks exit the freezer, they are culled for breading uniformity, packaged into either consu[mer] packages (8 oz. or 22 oz.) or large food service cartons (10 pounds), then labeled and passed through a m[etal] detector. Packages are cased, palletized, and stored in the freezer at -10°F. Product is shipped on fre[ezer] trucks to retail or food service distribution centers.

EXAMPLE 11-3

FOR ILLUSTRATIVE PURPOSES ONLY

Breaded Fish Sticks
Flow Diagram

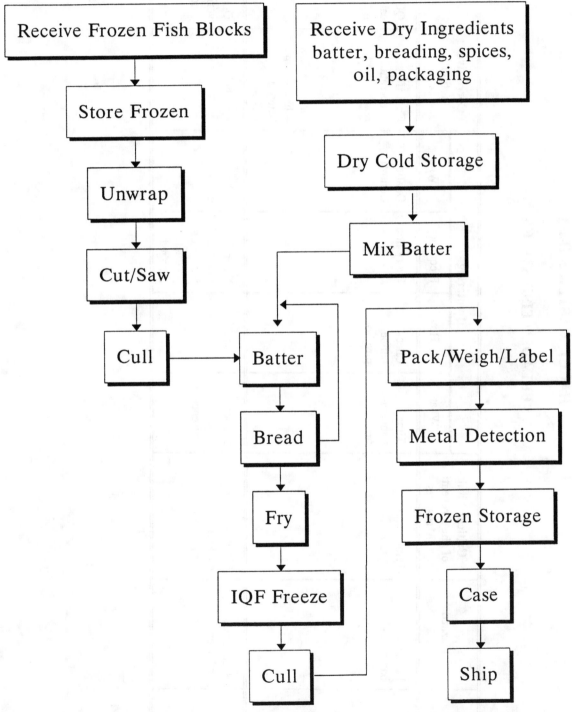

Figure 11-5. FLOW DIAGRAM FOR BREADED FISH STICKS

EXAMPLE 11-3

FOR ILLUSTRATIVE PURPOSES ONLY

HACCP MASTER SHEET

BREADED FISH STICKS

Critical Control Point (CCP)	Hazard	Critical Limits of the Preventive Measures	Monitoring				Corrective Action	Records	Verification
			What	How	Frequency	Who			
Batter CCP1(B)	Bacterial pathogens	Temperature <45°F	Time, temperature	Check temperature hold tank	Every hour	Quality Control	Cool if temperature reaches 50°F If batter temp. is over 45°F for more than 4 hrs, dump batter & reclean	QC log, calibration records	Check records daily, calibrate thermometer weekly. Audit.
Package label CCP2(B)	Bacterial pathogens	Label with proper cooking instructions	Proper label presence	Visual	When labels are loaded or product is changed.	Label operator	Re-label	Operators log, verification records	Q.C. checks for appropriate label hourly.

HACCP MASTER SHEET
BREADED FISH STICKS

Critical Control Point (CCP)	Hazard	Critical Limits of the Preventive Measures	Monitoring				Corrective Action	Records	Verification
			What	How	Frequency	Who			
Metal detector CCP3(P)	Metal	Operable metal detector	Ferrous and non-ferrous metal	Automatic screening	Continual	Labeling operator checks to ensure detector is on	Line operator evaluates rejects hourly. Excessive reject rate forces line shutdown.

Q.C. prepares deviation report, determines source of metal, and corrects it. Contact maintenance if necessary - fix detector.

If detector is not on, or fails sensitivity check, all product since last acceptable check is held and rechecked for metal. Operator pre-pares deviation report, notify Q.C./Q.A. | Operator log indicating time of metal detector sensitivity checks and results.

Deviation reports indicating lot and time when metal was detected or detector failed, results of evaluation, and disposition of product.

Metal detector maintenance and calibration records. | Run test material with metal of appropriate size to check sensitivity hourly (label operator).

Supervisor reviews operator logs daily, deviation reports weekly. |

SECTION IV

MANAGING HACCP PROGRAMS

Organizing and Managing HACCP Programs

The Relationship of HACCP to CGMPs and Sanitation

CHAPTER 12

ORGANIZING AND MANAGING HACCP PROGRAMS

by K. E. Stevenson and Dane T. Bernard

INTRODUCTION

The preceding sections of this manual have explored the concept of HACCP and provided detailed information on the development of a HACCP Plan. If a company decides to use HACCP as the system for assuring the safety of its products, then the HACCP Program becomes an integral part of the company's operations. Since HACCP represents a structured approach to control the safety of food products, it must be organized and managed in a manner which assures that the HACCP system is operating correctly and that it will be sustained and maintained appropriately for the foreseeable future.

ORGANIZING A HACCP PROGRAM

The manner in which HACCP functions are organized within a company varies considerably due to the variability of internal organizations and responsibilities of groups which may be present. However, the HACCP System is normally associated with the Quality Assurance (QA) group, or a similar group which formerly was responsible for the food safety functions in the company.

After developing a HACCP Plan, most companies will find that their current operations have controlled the critical factors associated with a product and process. However, it would be a mistake to assume that because a company now takes steps consistent with a HACCP Plan, that it is already employing HACCP. Deficiencies in the HACCP System are most often detected in two areas: (a) documentation of the HACCP Plan (insufficient "background" documentation related to decision-making, and inadequate documentation of actual processes), and (b) management of the HACCP Program (failure to ensure that a comprehensive system is in place to yield safe products, and inadequate review mechanisms to prove that the HACCP Plan is being applied correctly).

Management Commitment

In order for HACCP to succeed within a company, there must be a clear commitment to food safety and HACCP concepts. Like many other systems and organizations, success hinges on management commitment, detailed planning, appropriate resources and employee empowerment. Thus, a key step in initiating work on a HACCP Program is a corporate commitment to HACCP (and food safety) which is communicated throughout the company.

A simple statement presenting the corporate policy with respect to HACCP is an important step in documenting management commitment. This represents a powerful tool in communicating the importance of HACCP to every employee in the organization. In addition, specific objectives and general implementation schedules should be established by management in cooperation with the HACCP Team.

HACCP Coordinator and HACCP Team

The responsibility and authority given to the HACCP Coordinator and the HACCP Team represent another key way in which to communicate the importance of HACCP. The HACCP Coordinator must be chosen carefully, since this individual will be the "champion" of HACCP within the company. This individual, in conjunction with management, should select the personnel who will become members of the multi-disciplinary HACCP team and provide the leadership and guidance for development of the company's HACCP Plan. Also, management should assure that personnel receive essential training in HACCP, and that resources are available to obtain appropriate outside expertise, if necessary. (See Chapter 7 for additional information related to the HACCP Coordinator and HACCP Team.)

A Strategy for Developing a HACCP Plan

Once assignments to the HACCP team have been made, there is a tendency to become overwhelmed by the complexity of the operations and the tremendous amount of information and documentation which must be generated in the development of a comprehensive HACCP Plan. As in many complex jobs, careful planning, appointment of subgroups and assignment of small tasks is essential.

A suitable approach to developing a HACCP Program, is to begin by developing a "model" HACCP Plan for one specific product and process. The HACCP Team should gather appropriate information and gain knowledge of the specified product and process so the HACCP Principles can be applied in an appropriate manner. This task involves all of the details associated with describing the food, its intended use and distribution, and developing and verifying a flow diagram which describes the process.

Once developed, this HACCP Plan can serve as the model for the development of additional HACCP plans for other products. The experience gained and the procedures used to develop this first HACCP Plan will facilitate the development of HACCP Plans for additional products and product groups. Furthermore, a single HACCP Plan may be used for a group of very similar products, provided that the HACCP Team assures that this Plan is appropriate for each of the products and their processes.

Note: The process of developing HACCP Plans for additional products and processes can be expedited by the formation of specific product/process teams to assist in their preparation. These teams would have the responsibility of developing HACCP Plans, based on the design and procedures used by the HACCP Team to develop the initial HACCP Plan. Obviously, at least one key member of each product/process team should be a member of the HACCP Team.

Implementing A HACCP Plan

HACCP Programs are no different than other management programs; some unforeseen problems will be encountered during implementation! Thus, a trial period should be used to allow employees to become familiar with the HACCP Plan and to attempt to discover any weaknesses or significant problems. During the trial period, the HACCP Plan may undergo relatively constant review, evaluation, and revision. However, a formal review should be scheduled to specifically evaluate the current system and to recommend revisions.

Arrangements should be made to provide appropriate training at all levels. This should include a general overview, so that all employees understand the general concept of HACCP and the HACCP policy and objectives, as well as, specific training associated with individual jobs and tasks.

MANAGING A HACCP PROGRAM

HACCP Programs are not automatic. In order to succeed, they need appropriate support and management systems. Some suggestions related to management of HACCP systems are included in the remainder of this Chapter.

Coordination of Food Safety Operations

One person should have overall responsibility for the food safety system (HACCP) within a company. This responsibility extends to input, review, and approval of the documentation of the HACCP system. This individual also should be able to assure that the HACCP Team has access to the variety of information which they will require to conduct their assignments. In large companies, there may be a separate individual who is responsible (from an operations standpoint) for every employee who is assigned a HACCP-Plan task associated with the specific product/process.

Regardless of whether or not you are working in a large company, every individual assigned to a HACCP-related task should receive appropriate (written) instructions and descriptions of their responsibilities and tasks. Clarification of reporting structures and the relationships of the various groups involved is imperative. Since food safety issues are preeminent, HACCP issues must take precedence over quality and production issues.

Systems for Evaluating New Products

Once HACCP Plans have been developed for all of the products being produced in an organization, there must be a structure established for evaluating new products and processes. In most instances, individuals working in product/process development areas do not have extensive training in food safety. Thus, it is imperative that a system be established to facilitate the evaluation of new products and processes with respect to food safety.

This system of evaluating new products and processes prior to scale-up and commercialization can be highly beneficial. First, it can provide an early indication of any food safety-related problems, thereby saving time and money. Second, once such a system is in place, product development employees will be more cognizant of food safety considerations when they are designing new products and processes.

Other safeguards can also be implemented as part of the overall management of the HACCP system. When a plant is currently operating with HACCP Plans in place, production of a new product should not begin until a HACCP Plan has been developed for the new product/process.

Systems for Evaluating Product/Process Changes

The HACCP concept demands that a mechanism be established to evaluate any proposed changes related to a product or process. The decision concerning whether or not a change in the product/process is significant--with respect to food safety--should be made by the HACCP Team or another group specifically appointed for that task. A mandatory evaluation process guarantees that a systematic evaluation will be made of any changes in the process or product, thus, assuring that any changes or revisions which might affect food safety will be thoroughly investigated prior to their implementation. A policy should also be in effect that prohibits changes from being made without such an evaluation.

Day-to-Day Management

Routine management of the HACCP Plan is facilitated by the requirements associated with monitoring and the daily review of records associated with Critical Control Points (CCPs). Designing relevant assignments and records streamlines this task. Furthermore, documentation of reporting responsibilities and appropriate corrective actions also clarifies actions to be taken and helps assure that the correct individuals are notified immediately when a problem has been discovered.

One of the more visible benefits of a HACCP system is the fact that management can now receive daily reports related to food safety. In addition, this type of activity can frequently spot trends which can trigger adjustments to a process or operation *before* it results in a problem.

Periodic Evaluation and Revision

Verification procedures assure that the HACCP system will be evaluated and revised on a periodic basis. In some cases, a problem may occur that is not recognized on a day-to-day basis, either due to an imperceptibly gradual change or because of a failure to comprehend a potential problem. Therefore, the periodic evaluations are invaluable in evaluating the HACCP system. In addition to evaluating long-term trends, these evaluations also are used to determine if any changes need to be made in the HACCP Plan procedures or documentation. A written report of the findings and recommendations of this verification procedure should be sent to management, and the report also should become a part of the documentation in the HACCP Plan Master File.

SUMMARY

The HACCP concept is intended to provide a systematic, structured approach to assuring the safety of food products. However, there is no blueprint or universal formula for putting together the specific details of a HACCP Plan. The strength of a HACCP program is in providing a system which a company can use effectively to organize and manage the safety of the products which are produced.

CHAPTER 13

THE RELATIONSHIP OF HACCP TO CGMPs AND SANITATION

by Allen M. Katsuyama and Bonnie J. Humm

INTRODUCTION

Contrary to popular perception, sanitation is not limited to the cleaning of equipment. Although clean equipment and a clean environment are essential for producing safe foods, equally important are personnel practices, plant facilities, equipment and operations designed to prevent contamination, pest controls, and warehousing practices. All of these considerations should be addressed in a comprehensive sanitation program designed to comply with existing regulations. Such programs should be viewed as essential **prerequisites** for an operation which is considering development of a HACCP system. In addition, many biological, chemical, and physical hazards can be avoided by taking the precautions outlined in the following regulatory requirements and guidelines:

21 *CFR* 110	Current Good Manufacturing Practices (CGMPs)
9 *CFR* 308	Sanitation for Meat Products
9 *CFR* 381	Subpart H: Sanitation for Poultry Products

CGMP/SANITATION REQUIREMENTS

The Current Good Manufacturing Practices regulations, which were revised in 1986, were promulgated by the Food and Drug Administration to provide criteria for complying with provisions of the Federal Food, Drug, and Cosmetic Act (FD&C Act) requiring that all human foods be free from adulteration. Since acts resulting in food adulteration, as defined in the FD&C Act, include aesthetic and economic considerations, several specific details of the CGMP regulations do not deal directly with food safety. However, many of the requirements have some direct or indirect influence on the biological, chemical, or physical safety of the finished products.

The CGMP regulations are divided into several subparts, each containing detailed requirements pertaining to various operations or groups of operations in food processing facilities. Emphasis is placed on the prevention of product contamination from direct and indirect sources. The sanitation regulations for meat and poultry products, promulgated by the U.S. Department of Agriculture (USDA), contain identical or similar requirements.

General Provisions

Several terms used in the regulations are defined in order to minimize misunderstanding. Included among the terms in the CGMPs are "critical control points" and "quality control operations."

This subpart also contains a summary of responsibilities imposed on plant management regarding plant personnel. Criteria for disease control, cleanliness (personal hygiene and dress codes), education and training are articulated. These requirements are designed to prevent the spread of disease from worker to worker, from workers to the food processing area and from workers to the food itself. Also included is the requirement that a competent supervisory person be assigned the responsibility for assuring compliance by all personnel.

Buildings and Facilities

The section covering "Plant and Grounds" describes the general principles of plant design and construction necessary to protect food from insanitary conditions. The methods for adequate maintenance of grounds are enumerated. To reduce food contamination, several design mechanisms for the separation of various operations, including outdoor bulk fermentation vessels, are recommended. Adequate working space, lighting, and ventilation are required.

The "Sanitary Operations" section establishes basic rules for food plant sanitation. The rules describe general requirements for (a) maintenance of physical facilities (building and fixtures), (b) pest controls, (c) cleaning and sanitizing of equipment and utensils, and (d) storage and handling of cleaned equipment and utensils. The proper use and storage of cleaning chemicals, sanitizing agents, and pesticides are emphasized.

Minimum requirements for sanitary facilities and accommodations are described under "Sanitary Facilities and Controls." Included in this section are requirements for (a) water used for various purposes, (b) plumbing for both water and wastewater, including a rule specifically prohibiting backflows and cross-connections, (c) sewage disposal, (d) toilet facilities, (e) hand-washing facilities and supplies, and (f) rubbish and offal disposal.

Equipment

The "Equipment and Utensils" section describes general principles of design, construction, and maintenance of processing equipment and utensils. Cleanability is emphasized. Since precluding microbial contamination is crucial, requirements for equipment used to control or prevent growth of microorganisms are listed. This includes cooling equipment (freezers and cold storage) and other instruments/devices for measuring/controlling pH, acidity, water activity, etc. Requirements for compressed air and other gases used in food processing are also set forth. The use of polychlorinated biphenyls (PCBs) has been virtually eliminated from food and feed processing plants.

Production and Process Controls

The several sections and subsections in this subpart contain the most detailed requirements. There are rules to assure the suitability of raw materials and ingredients, to maintain the integrity of processed foods, and to protect finished foods from deterioration.

The "Process and Controls" section requires that all operations involving foods "be conducted in accordance with adequate sanitation principles," and that plant sanitation "be under the supervision of one or more individuals assigned responsibility for this function." Quality control programs, including appropriate testing procedures where necessary, are required to insure compliance with sanitation principles and with the FD&C Act. This requirement is based on FDA's experience in conducting thousands of establishment inspections annually. The agency has found that most companies have quality control procedures which help insure that raw materials and finished products are fit for food, packaging materials are safe and suitable, and all are in compliance with the Act.

The subsection dealing with raw materials and ingredients contains descriptions of required methods and procedures, including inspection, segregation, and washing or cleaning, to insure the cleanliness and fitness of these materials. These materials must not contain levels of microorganisms that can produce food poisoning or other disease in humans or they must be properly treated to destroy such microorganisms. Compliance with regulations on natural toxins and extraneous materials is stressed.

The need to protect foods from contamination is stressed in the subsection covering processing operations. Required conditions are listed for holding foods which support the rapid growth of microorganisms which cause foodborne disease, or those that can spoil foods as a result of multiplication. Requirements to insure freedom from adulteration are stated for several types of foods (batters and other preparations, intermediate moisture and dehydrated foods, and acidified foods), and for various operations (mechanical processing steps, heat blanching, and filling/packaging).

Warehousing and Distribution

This section requires that storage and transportation of finished foods be conducted under conditions preventing physical, chemical, and microbiological contamination. Although permanently legible code marks on all food packages are no longer generally required, the use of printed codes is highly recommended in the event that a recall becomes necessary to remove products from distribution channels. FDA has promulgated guidelines for policies and procedures of the agency and steps that firms should follow during product recalls (21 *CFR* 7, Subpart C).

Defect Action Levels

This section addresses "natural or unavoidable defects" which occur in some foods, such as raw agricultural commodities, and carry through to the finished product, but which present no human health hazard. The current level of defects permitted is based largely on the industry's ability to reduce the levels occurring in the raw product through good manufacturing practices. The "action" levels represent the limits at or above which FDA may take legal action to remove the commodities from the consumer market.

These defect action levels cannot be used as an excuse for poor manufacturing practices. FDA clearly advises that failure to operate under good manufacturing practices will leave a firm liable to legal sanctions even though products might contain natural and unavoidable defects at levels lower than the currently established action levels. Additionally, the mixing of food with defects above the action level and foods with defects below the action level to produce a product with an acceptable defect level is prohibited. The "blended" lot is unlawful regardless of the defect level in the finished product.

CGMP COMPLIANCE

Since the CGMP regulations and the USDA sanitation requirements address some biological, chemical, and physical hazards associated with the production of foods, compliance with specific portions of these regulations and requirements may be included as part of the HACCP program. However, assuring compliance with any critical control point (CCP) requires documentation. Therefore, as with any other CCP, any CGMP requirement included in the HACCP program must be explained in detail. Regularly scheduled inspections must be conducted to document and verify that all CCPs are in control.

The plant sanitation or CGMP compliance program should consist of formal, written plans and procedures that are collated in a sanitation manual. The plan or procedure for each aspect of the program must be sufficiently detailed to insure consistent compliance with the pertinent requirement. The following discussions address some of the CGMP compliance points.

Employee Practices

Food plant employee practices can result in food contamination. Employees contaminate foods microbiologically by touching foods, tools, implements, and food-contact surfaces with unclean hands, hair, and soiled clothing. Personal effects (jewelry, pens, hair pins, etc.) are often found in foods, as are other employee generated foreign objects, such as candy and gum wrappers, cigarette butts, coins, and the like. Also, the improper or negligent use of chemicals by employees may result in chemical contamination of foods. Thus, employee practices must be effectively addressed to prevent unacceptable levels of contamination.

Since the operative limit for employee practices is strict adherence to proper hygienic practices, effective employee training and adequate supervision are essential. A strong company policy must be written and enforced. Every new employee must be informed of GMPs and their role in the production of safe foods. Responsibility for continuous monitoring of hygienic practices must be assigned to area supervisors and regularly verified and documented through scheduled inspections and audits. Violations must be corrected as soon as possible; flagrant infractions which may result in an unacceptable food safety hazard must be dealt with appropriately and may even warrant immediate dismissal. In a labor union environment, getting union leaders to understand and cooperate in CGMP compliance is a key factor for a successful program.

Mandatory employee meetings should be scheduled regularly to review and improve present practices. Involving employees in this decision process will allow them to "buy into" the program and offer them the opportunity of making meaningful suggestions for improving the program, including decisions involving disciplinary actions.

Pest Controls

Sanitation programs must include detailed procedures for controlling pests. Although some may argue that contamination of foods by pests is limited to aesthetic and economic adulteration, many pests, such as cockroaches and birds, can transmit microbiological hazards to food processing environments or foods. In addition, some foods, such as grains and nuts, are susceptible to mold and the subsequent development of mycotoxins following insect damage.

The pest control program must be comprehensive and should be based on the integrated pest management (IPM) philosophy. Rather than relying on the use of pesticide chemicals to eradicate a pest infestation, an IPM program is largely preventive, employing physical and mechanical control measures in addition to the use of chemicals. Therefore, IPM programs are safer and much more effective than relying on just one method.

IPM dictates the initial use of physical methods to exclude pests from the food establishment. The facility must be tightly constructed to deny access. Harborage and attractants around the plant must be eliminated. Maintaining sanitary conditions and eliminating sources of food, water, and shelter will reduce the number of pests around and in the plant.

Primary reliance should be placed on mechanical devices to eliminate pests that enter the facility. Rodent traps and insect electrocutors are examples of excellent, effective mechanical devices when properly installed and maintained.

Despite effective physical barriers and mechanical devices, periodic use of pesticide chemicals will be required. However, the need for chemicals should be minimal and their use must be employed with proper caution. The IPM program should be developed with personnel who are responsible for the implementation of the program, either a trained employee or an outside pest control operator (PCO). Records associated with the IPM program should include, but not be limited to, the following:

- Map of rodent stations, bait stations, and insect electrocutors.
- Maintenance schedule for rodent stations, bait stations, and insect electrocutors.
- List and inventory of all pesticides used for the program, including a copy of all labels.
- SOPs for pesticide application by in-house personnel.
- Copies of all reports by an outside PCO, listing insects and/or rodents found, areas of pest activity, application of any pesticides (name of chemical and amount applied).
- Reports of all in-house pest control inspections with "corrective actions" listed.

- Reports of all problems with the physical facility, or equipment out of compliance with the plant sanitation program, with "corrective action taken" and "by whom" clearly delineated.

The pest control records cited above are required by regulations and serve as part of the essential documentation for a sanitation program.

Cleaning Procedures

A comprehensive sanitation program includes detailed cleaning procedures for all equipment and all production lines in the food processing facility. Procedural details will include a list of chemicals used, the methods of application, contact times, rinsing procedures, sanitizing requirements, and whether a piece of equipment can be cleaned-in-place (CIP) or must be disassembled and cleaned-out-of-place (COP). Special attention must be placed on equipment used for handling foods which receive no subsequent heat treatment (e.g. a meat-slicing operation for ready-to-eat refrigerated foods) since any microorganisms contaminating such foods will not be destroyed.

The sanitation of a specific piece of equipment may be identified as a Critical Control Point. In such instances, the adequacy of the cleaning and sanitizing procedures must be monitored. Daily post-cleanup inspections, as well as daily pre-operational inspections, with appropriate inspection forms can be used to monitor and document cleanup efficiencies. Corrective action must be taken immediately whenever deficiencies from acceptable performance are observed. Before the operations are resumed, a responsible individual should be given the authority for a final "go" or "no go" decision based upon the adequacy of the cleanup.

MANAGING FOR SUCCESS

The management of sanitation programs, as well as HACCP systems, involves a pro-active approach to product safety and the participation of employees at every level in the decision-making process. Employee education is the foundation of sanitation and CGMP programs, and management must exercise a strong commitment of money, personnel, and materials to accomplish the task. Since personnel changes occur frequently in the industry, the educational process must be an on-going program requiring considerable commitment from all those involved. Guidelines that can be used to successfully implement and manage CGMP and sanitation compliance programs, as well as the HACCP system, include the following:

- Obtain the full commitment of upper management.
- Allocate adequate funds to support the program.
- Provide effective leadership and discipline.
- Provide written procedures for all programs.
- Initiate training programs for employees based on employee participation in decision making.
- Encourage the full participation and cooperation of all employees.

- Develop monitoring procedures and recordkeeping forms to document compliance with written procedures.
- Implement verification procedures to determine program effectiveness.
- Develop "feed-back" procedures for continuous improvement.

SUMMARY

Contrary to popular perception, sanitation involves much more than just keeping equipment clean. It encompasses all activities and responsibilities for preventing product adulteration. Sanitation also includes the implementation of appropriate actions to prevent the occurrence of some hazards that can harm consumers.

FDA's current good manufacturing practices regulations and USDA's sanitation requirements serve as the basis for food plant sanitation programs. Effective sanitation programs include (a) the delineation of good employee practices, (b) the development of detailed procedures for cleaning and maintaining the facility and equipment, (c) a comprehensive pest control program, (d) inspections of incoming materials and of the total operation, (e) monitoring of critical physical factors, and (f) storing and distributing finished products in a manner that insures their safety. Therefore, good hygienic practices and compliance with CGMPs are essential prerequisite programs for a well designed HACCP system.

SECTION V

HACCP AND REGULATIONS

HACCP and The Regulatory Agencies

Recalls: Regulations/Company Organization

CHAPTER 14

HACCP AND THE REGULATORY AGENCIES

by Lloyd R. Hontz and Dane T. Bernard

FEDERAL FOOD INSPECTION OVERVIEW

INTRODUCTION

Other than several HACCP-based GMP regulations, use of HACCP within the food industry to date has been voluntary. One early government advisory group addressing the topic of food safety noted that the fullest potential for HACCP-based protection of our food supply may not be realized until it is mandated by regulations. While not all would agree with this statement, efforts have begun by food regulatory agencies to institute HACCP on a mandatory basis. Despite efforts by many to assure harmony of requirements, the degree to which HACCP will eventually be adopted by specific segments of the food industry will be influenced by the Federal regulatory agency having jurisdiction. In addition to the situation in the U.S., HACCP has taken on international importance as a potential tool to judge the acceptability of food products in trade. For these reasons, food firms are advised to seriously consider developing HACCP plans for the products they manufacture.

Currently, there are two primary U. S. Federal agencies with food safety missions. They are the U.S. Department of Agriculture's Food Safety and Inspection Service (FSIS) and the Food and Drug Administration (FDA) of the U.S. Department of Health and Human Services.

USDA/FSIS INSPECTION OF MEAT AND POULTRY PRODUCTS

As mandated by the Federal Meat Inspection Act and the Poultry Products Inspection Act, FSIS has jurisdiction over the production of the country's food products which contain meat or poultry. This encompasses animal slaughter as well as "further processing" operations. In general, FSIS has oversight for all food products containing more than 2-3 percent meat or poultry.

FDA REGULATORY JURISDICTION OVER FOOD PRODUCTS

Under the Federal Food, Drug and Cosmetic Act, FDA's mandate includes virtually all food products moving in interstate commerce, other than those inspected by FSIS.

NMFS/FDA VOLUNTARY SEAFOOD INSPECTION PROGRAM

Certain categories of food are subject to additional regulatory oversight. For example, the Department of Commerce's National Marine Fisheries Service (NMFS) has developed a voluntary, HACCP-based, "fee for service" inspection program for seafood. Through this program, processors may request NMFS approval of quality assurance and food safety programs used by the firm. If approved, the firm may then affix a "seal" to products manufactured under this continuous federal inspection program.

PASTEURIZED MILK ORDINANCE

Producers of milk and milk products must also comply with state regulations which are typically based on a model ordinance termed the Grade A Pasteurized Milk Ordinance (PMO). While this document is based on a HACCP-type hazard assessment, it is quite prescriptive in nature. There are a number of major differences, and many other less significant variations, when comparing the requirements in the PMO with other food safety regulations pertaining to the same products.

PROPOSED APPLICATION OF HACCP PRINCIPLES BY REGULATORY AGENCIES

As noted above, significant efforts are now underway by both FDA and USDA to shift the focus of regulatory oversight to a HACCP-based inspection system. Both agencies plan to base their initiatives on application of the National Advisory Committee on Microbiological Criteria for Food's HACCP document adopted in 1992 (chapter two in this manual).

On January 28, 1994, FDA published a proposed rule "Proposal to Establish Procedures for the Safe Processing and Importing of Fish and Fishery Products" (59 FR 4132). If finalized, this rule will mandate that each processor or importer of seafood products develop and implement HACCP plans to help assure the safety of these products. The comment period for this proposal closed in May, 1994. As of this writing, the agency is still reviewing comments and is not expected to publish a final rule until late in 1995.

In addition, FDA published an Advance Notice of Proposed Rulemaking (ANPR), "Development of Hazard Analysis and Critical Control Points for the Food Industry" (59 FR 39888). In this document the agency asked over one hundred rhetorical questions regarding if, when, and how it should apply HACCP to industry segments other than seafood. The comment period has closed on the ANPR and the agency is considering the comments it received in response to this inquiry along with those from the seafood proposal.

On February 3, 1995, the Food Safety and Inspection Service of USDA issued a proposed rule, "Pathogen Reduction; Hazard Analysis and Critical Control Point (HACCP) Systems" (60 FR 6774) which would mandate HACCP for the meat and poultry industry. The proposal covers not only slaughter establishments but "further processing" establishments as well. This sizeable proposal also includes a variety of near term requirements for pathogen reduction. The comment period closes on June 5, 1995. The agency must then review all comments and respond to each before a final rule is published, possibly by the end of 1995.

It is always important for food processors to stay abreast of the regulations and proposed rules applicable to their operations. At this time processors have an opportunity to influence the future of food safety assurance by providing appropriate input into the agencies. Since the mechanism(s) for incorporation of HACCP into food production and inspection activities is still under development, food processors can participate in the process and influence the manner in which HACCP is eventually applied.

USE OF HACCP CONCEPTS IN CURRENT FOOD REGULATIONS

FDA LOW-ACID CANNED FOOD REGULATIONS

As a result of several outbreaks of botulism from consumption of improperly processed canned food products in the early 1970's, the National Food Processors Association (then the National Canners Association) petitioned FDA to adopt a comprehensive regulation which would describe good manufacturing practices for canned foods. The resulting regulations, currently found in 21 CFR 113, "Thermally Processed Low-Acid Foods Packaged in Hermetically Sealed Containers," are frequently cited as the first regulatory use of HACCP concepts in the food industry. Indeed, without actually using HACCP terminology, these regulations identify the critical factors (analogous to CCPs), specify certain (critical) limits or require that proper controls be determined by processing authorities, and require manufacturers to develop and maintain records to indicate that the critical factors are under control. Other elements of HACCP, such as predetermination of steps to take when critical limits are not met (corrective actions for process deviations), and periodic inspection by FDA personnel to verify that the proper records are being kept, are important features of these regulations. FDA later promulgated similarly styled regulations for the production of acidified food products (21 CFR 114, Acidified Foods). Together these regulations have been very successful in helping to assure the safety of canned food products.

FSIS CANNED FOOD REGULATIONS FOR MEAT AND POULTRY

In 1986, after years of industry encouragement, FSIS revamped its requirements for production of canned meat and poultry products. The random system of fragmented regulations, inspection manual requirements, and Agency bulletins, directives and notices were replaced by a comprehensive set of canning regulations patterned after the FDA model. These regulations similarly are based upon HACCP concepts.

INTERNATIONAL TRADE - HACCP AND CODEX ALIMENTARIUS

The Codex Alimentarius Commission's Committee on Food Hygiene has also played an active role in formulation and encouragement of HACCP as an international tool to assure the production of safe food products. At the 25th session of the Codex Committee on Food Hygiene (Oct-Nov, 1991) the document "General HACCP Definitions and Procedures for Use by Codex" was introduced. The committee agreed that HACCP should be incorporated into Codex Codes of Practice and the General Principles. A working group was appointed to further develop the document, and in only two years the document was completed and presented to the Codex Alimentarius Commission which adopted it as final action. Codex involvement brings even greater potential for international harmonization and understanding of the HACCP principles.

During the 1992 Quadrilateral Discussions on Food Safety, regulatory representatives from Australia, Canada, New Zealand and the United States reviewed the current status of HACCP within their respective countries and discussed the role of regulators in the application of HACCP principles. Discussions between these countries continues to drive progress in international utilization of HACCP as a food safety management tool of international importance.

Other International bodies including the Food and Agriculture Organization (FAO) of the United Nations and the World Health Organization (WHO) have held international consultations on the topic of HACCP. The purpose is to further the international understanding of HACCP and to help develop training materials and expertise which can be made available to all nations, including developing countries.

FUTURE USES OF HACCP

The precise manner in which HACCP will be applied to various segments of the food industry is not yet clear. What is clear, however, is that many firms are voluntarily implementing HACCP programs on their own initiative. In fact, many firms are requiring their suppliers and co-packers to develop and implement HACCP plans as a requisite for doing business. The extent of this is a clear indication of the benefits that companies perceive from the use of the best system available today for assuring the safety of their products.

REFERENCES

FDA. 1995. Acidified foods. Title 21, *Code of Federal Regulations*, Part 114. U.S. Government Printing Office, Washington, D.C. (Issued annually)

FDA. 1995. Current good manufacturing practice in manufacturing, packing, or holding human food. Title 21, *Code of Federal Regulations*, Part 110. U.S. Government Printing Office, Washington, D.C. (Issued annually)

FDA. 1995. Thermally processed low-acid foods packaged in hermetically sealed containers. Title 21, *Code of Federal Regulations*, Part 113. U.S. Government Printing Office, Washington, D.C. (Issued annually)

NACMCF. 1992. Hazard analysis and critical control point system. *Intl. J. Food Microbiol.* 16:1.

NAS. 1985. *An Evaluation of the Role of Microbiological Criteria for Foods and Food Ingredients*. National Academy Press, Washington, D.C.

NMFS. 1987. *Plan of Operation - Model Seafood Surveillance Project*. NMFS, Office of Trade and Industry Services, National Seafood Inspection Laboratory, Pascagoula, MS.

USDA. 1995. Mandatory meat inspection. Title 9, *Code of Federal Regulations*, Parts 301-335. U.S. Government Printing Office, Washington, D.C. (Issued annually)

USDA. 1995. Mandatory poultry products inspection. Title 9, *Code of Federal Regulations*, Part 381. U.S. Government Printing Office, Washington, D.C. (Issued annually)

CHAPTER 15

RECALLS: REGULATIONS/COMPANY ORGANIZATION

by National Food Processors Association[1]

FEDERAL REGULATIONS RELATING TO RECALLS

Food manufacturers produce billions of containers each year, processing, packaging and distributing their products world-wide. Seldom is there a reason to question the safety of the tremendous variety of foods made routinely available to consumers, whether the products are refrigerated, frozen, canned, or dehydrated.

The primary reason for the food industry's outstanding safety record is the precautions and safeguards built into the manufacturing process. These precautions and safeguards have been developed jointly by the food industry and federal regulatory agencies. HACCP is an important tool for improving and maintaining the excellent safety performance of the food industry.

Legal responsibility for product safety and wholesomeness rests with the officers and employees of a food manufacturing company. The corporate officers bear the ultimate responsibility. It is in their best interest, and those of the company and its consumers, to have a system in place that assures the smoothest possible handling of any product withdrawal, stock recovery or recall that may become necessary. Regulations related to production of low-acid foods require companies have such a "recall system." All food companies should develop a recall system in case a HACCP-plan-deviation food is in distribution, and found to be a potential health hazard.

Failure to remove contaminated or illegal products from the market can result in damaging publicity, consumer complaints, product liability suits, civil and criminal lawsuits by federal or state government agencies and damage to the company's reputation. When confronted with a potential recall situation, a company must decide whether to withdraw or recall its product, and how to do it, on the basis of its own business, and legal or regulatory considerations. In addition, each food manufacturer must be prepared to deal with a product emergency. The key to being prepared is to develop a Recall Team and a Recall Plan that can rapidly be put into action when the crisis hits.

Guidelines on policy, procedures, and industry responsibility for recalls under jurisdiction of the Food and Drug Administration are listed in 21 CFR 7.40 through 7.49. This regulation was published on June 16, 1978. (Note: There are special rules pertaining to recalls of infant formulas). Procedures for recall of meat and poultry products subject to jurisdiction of the Food Safety and Inspection Service of the U.S. Department of Agriculture are spelled out in the FSIS Directive 8080.1, Revision 2, dated November 3, 1992.

---

[1]This material was adapted by Cleve B. Denny from *NFPA Bulletin 34-L: Successfully Managing Product Recalls & Withdrawals.*

Neither the Food and Drug Administration nor the U.S. Department of Agriculture has authority to require a recall without obtaining a court order. In practice, the FDA and USDA's Food Safety and Inspection Service (FSIS) use their persuasive powers to convince company officials that a recall is needed. The action is sometimes viewed as an alternative to an agency seizure or detention action to remove or otherwise correct adulterated or misbranded products.

Food processors are free to present information to FDA and USDA that might cause the agencies to reevaluate a given situation. But the company must seriously consider the possible adverse consequences of refusing a request for a recall action by FDA or FSIS. The agency may on its own initiative issue a statement to news media pointing to any deficiency in a company's products that it believes poses an imminent danger to health or could result in gross deception of consumers. Adverse publicity from a subsequent press release or news broadcast can have a devastating impact on any food company, and can severely damage the reputation of its products in the marketplace.

If a company refuses to undertake a recall requested by one of the agencies, or if government officials consider the company's action to be ineffective, the agency can seek a court order permitting it to seize or condemn products that are not in compliance with statutory requirements. Most states also have the authority to seize or hold adulterated or contaminated food products.

RECALL CLASSIFICATIONS

If FDA or FSIS undertakes to have a company recall a product, the recall will be given one of the following classifications:

Class I Recall: Used in situations where there is a reasonable probability that use of or exposure to a product will cause serious adverse health consequences or death.

Class II Recall: For cases in which use of or exposure to a product may cause temporary or medically reversible adverse health consequences or where the probability of serious adverse health consequences is remote.

Class III Recall: Employed when use of or exposure to the product is not likely to cause adverse health consequences.

In addition to recalls, these actions are available to a company that decides to regain control of product already in distribution:

Market Withdrawal: Used when there is a minor violation that is not subject to legal action by FDA or FSIS, or when the company wishes for other reasons to retrieve product from distribution, e.g. the product does not meet the company's internal specifications.

Stock Recovery: Employed in recovering product that has not been placed in retail distribution channels but is still under the manufacturer's direct control on its own premises or in warehouses from which the company can assure there will be no distribution.

Correction: Covers steps taken to repair, modify, relabel or otherwise adjust a product so that it may remain in distribution, or, destruction of a product with regulatory concurrence.

HEALTH HAZARD EVALUATION AND CLASSIFICATION

If a recall is to be made, the agency, not the company, classifies the recall. Once notified of a product defect, FDA and FSIS will evaluate its potential seriousness. FDA's guidelines state that an ad hoc committee of FDA scientists will evaluate the health hazards presented by a product being considered for recall, taking at least these factors into account:
1. Whether disease or injuries have resulted from use of the product.
2. Whether any existing conditions could contribute to a clinical situation that could expose humans or animals to a health hazard.
3. The degree of hazard to various population segments likely to be exposed to the product, particularly to those people at greatest risk.
4. The seriousness of the health hazard to which populations at risk would be exposed.
5. The likelihood of occurrence of the hazard.
6. An assessment of the consequences (immediate and long-term) of the hazard.

Based on the determination of the committee, FDA assigns a recall a classification (Class I, Class II, or Class III) to indicate the relative degree of health hazard of the product being recalled or considered for recall.

For products regulated by USDA, the health hazard evaluation is conducted by a team of FSIS experts in cooperation with other individuals or agencies as deemed necessary. The FSIS evaluation includes, at a minimum, consideration of the nature of the violation or defect, and items 1, 5 and 6 shown above.

RECALL STRATEGY

Once a company concludes that it must remove a product from distribution, it should develop a strategy for carrying out that task. This strategy must address these considerations:
1. Results of the health hazard evaluation process.
2. Codes, label names and other ways by which consumers can identify the product.
3. The degree to which the product's deficiency is obvious to the consumer or user.
4. The geographical area of distribution.
5. The extent to which the product remains under the control of the manufacturer or its distributors.
6. Whether the manufacturer will pick up the product or wants it shipped back and how payment will be made for the goods.
7. How the questionable product can be distinguished from other items so that good product continues to be made available to consumers.

AGENCY APPROVAL

The recall strategy developed by the company must be submitted to FDA or FSIS for approval, but the company need not wait for official approval to begin taking action or recover the product. In its review of the plan, even if that is done informally, the agency will recommend any changes it believes will make the plan more effective.

ELEMENTS OF A RECALL STRATEGY

The recall strategy will describe how deep the recall will be taken into the distribution chain, public warnings to be issued, and effectiveness checks.

Depth of Recall

The strategy will specify whether the recall is to be extended to the wholesale level, to the retail level, or if it is to be taken all the way to the consumer or user level.

Public Warning

A public warning is issued when the product being recalled presents a serious hazard to health (Class I). FSIS policy permits exemptions to the public notification requirements for Class I recalls under certain limited circumstances. Such a warning is used in urgent situations where other means of preventing consumption or use of the product would be inadequate. The manufacturer should develop a press release or statement and clear it with FDA or FSIS. In most cases, the manufacturer should issue the news release, going to news media outlets that will assure that consumers are alerted to the risk wherever the product was distributed. Sometimes, the agency will insist on issuing a public warning in order to maximize news coverage of the product defect. The manufacturer should insist on having input into any warning statement.

Effectiveness Checks

Effectiveness checks are required to verify that all consignees (at the recall depth specified by the strategy) have been notified about the recall or withdrawal and have taken appropriate action. The consignees may be contacted by personal visits, telephone, letters, telegrams or a combination of these methods. The manufacturer is usually responsible for conducting effectiveness checks. The responsible agency will help where necessary. The recall strategy should specify the method(s) to be used and the level of effectiveness checks to be conducted. These levels are recognized in FDA's recall policy and will be designated by FDA for each recall action:

- **Level A:** All consignees to be contacted.
- **Level B:** Some percentage of the total number of consignees to be contacted, with that percentage specified by the agency.
- **Level C:** Ten percent of the consignees to be contacted.
- **Level D:** Two percent of the consignees to be contacted.
- **Level E:** No effectiveness check to be made.

At FSIS, the number of effectiveness checks to be conducted is determined on a case-by-case basis in consideration of the recall level, health hazard, initial effectiveness review findings and the recalling firm's actions. In the event effectiveness reviews disclose recalled product remaining in commerce, FSIS may detain and seize product if the firm does not take prompt action following notification by FSIS of the problem.

FIRM INITIATED RECALL OR WITHDRAWAL

If a company decides to recall or withdraw a product from distribution because it believes the product is in violation of applicable laws or regulations, the company immediately should notify the local FDA district director or the FSIS regional office within 24 hours. The usual method of notification is a telephone call followed by written confirmation of details associated with the action. The company should provide the following information.

1. The product's identity.
2. The reason for removal or correction and details about when and how any deficiency was discovered.
3. An evaluation of the risk associated with consumption of the product, and how the evaluation was made (although the agency will make its own determination of risk).
4. The quantity of questionable product produced and the time span of that production.
5. An estimate of how much of the product is in distribution channels and how long it has been there.
6. Information about which distributors or customers received the product. If product has been diverted from one or more geographic areas to others, the recall can be considerably broader and more complicated.
7. A copy of any company correspondence with distributors, brokers or customers relating to recall strategy or actions, and a copy of any proposed news release.
8. Names, titles, and telephone numbers of company officials who will have a role in the recall action.

RECALLS REQUESTED BY AN AGENCY

The appropriate federal agency will ask a company to initiate a recall if it determines that a product has been distributed that poses a risk of illness or injury or gross consumer deception, that action is necessary to protect the public health and welfare, and if the manufacturer has not initiated its own recall action. The agency may notify the processor by letter, wire or oral communication of its request that a recall be initiated.

RECALL STATUS REPORT

The recalling firm is requested to file periodic recall status reports or effectiveness checks with the appropriate field office of the agency overseeing the action. Generally, the reporting interval will be two-four weeks, although the frequency of the reports depends upon agreements between the agency and the company.

Unless otherwise specified, the recall status report should contain the following information:

- Number of consignees notified of the recall, the date and method of notification.
- Number of consignees responding to the recall communication and the quantity of product each had on hand at the time it was received.
- The number of consignees that did not respond (the agency may request their identity).
- The quantity of product returned or held by each consignee.
- Estimated time frame for completion of the recall.

WEEKLY FDA ENFORCEMENT REPORT

Each week, FDA publishes a descriptive listing of recalls, whether requested by the FDA or initiated by the manufacturer. This report does not include information about a firm's product withdrawals or stock recoveries. However, corrections are sometimes reported. Occasionally, reports about recalls appear after a product has been completely withdrawn, corrected, and restocked.

The report, which includes other FDA regulatory actions such as seizures and filings of injunctions and prosecutions, is available to the news media. Unfortunately, the report occasionally triggers unfortunate publicity about a recall that already has been completed and may cause a second wave of media interest.

ENDING THE RECALL OR WITHDRAWAL

A recall will be concluded when FDA or FSIS determines that all reasonable efforts have been made to remove or correct the product in accordance with the recall strategy, and when it is clear that the product subject to the recall has been removed from distribution and proper disposition or correction has been made by the manufacturer. The agency will give the processor written notice that the recall is officially terminated.

A recalling company may request termination of its recall by submitting a written request to the appropriate FDA or FSIS regional office making the case that the recall has been effective, including the most recent recall status report and a description of what disposition has been made of the recalled product.

DO NOT destroy recalled product without offering to have an agency representative present. Keep good records of actions taken and the number of containers affected. Submit a written plan to the agency before doing anything with recalled product. The agency will inform the company if it objects to the plan.

COMPANY ACTIONS - ORGANIZING FOR A RECALL

Every food processor should develop a recall plan and an organizational structure that allows it to accomplish product withdrawals quickly and efficiently, as required by 21 CFR 108 and 9 CFR 318 and 381.

THE RECALL TEAM

Creating a Recall Team gives your company the means to prepare in advance to handle product emergencies. A good Recall Team, equipped with the right information, can assure that recalls and withdrawals are handled smoothly, with the least possible disruption to your operations. The Recall Team has these responsibilities:

- Recommend changes in operating procedures within the company that will lessen the possibility of having to withdraw defective products from the marketplace.
- If there is a suspected problem, assess the situation to determine whether or not the problem is real.
- Arrange for stock recovery or withdrawal if the product is not in violation.
- Develop the company's Recall Plan.
- Manage any correction or recall, including recovery, relabeling or other disposition of affected product, and payments to cover cost of the recall.
- Inform appropriate company employees and customers of any corrective actions undertaken.
- Coordinate with federal or state regulatory agencies and with trade association(s).
- Direct the company's recall activities.

The Recall Team should include representatives from production, purchasing, marketing, quality assurance, legal, sales, distribution, consumer affairs and public relations. Each of these disciplines is likely to be brought into play once a recall plan is put into action.

Develop, fill out, and maintain a recall organization chart (Figure 15-1) for your company. In deciding how to manage a recall or withdrawal, the recall team may need a variety of information, including:

- Facts about raw materials, ingredients and containers used in the defective product.
- Information about its own manufacturing operations, including thermal processing records, acidification records and container closure records required under 21 CFR 113, 21 CFR 114, 9 CFR 318 and 9 CFR 381.
- Information about any changes in equipment or equipment breakdowns or malfunctions.
- Quality assurance records, including cook checks.
- Facts about product on hand from the suspected code lot(s).
- Details of shipping dates, amounts and consignees.
- Consumer complaint records.
- Knowledge of press coverage of the recall.

Figure 15-1. Recall Organization Chart

CEO or President _____ Phone # _____
                                         Home Phone # _____

Recall Coordinator _____ Phone # _____
                                               Home Phone # _____

Alternate _____ Phone # _____
                                 Home Phone # _____

Distribution _____ Phone # _____
                                        Home Phone # _____

Alternate _____ Phone # _____
                                 Home Phone # _____

Production _____ Phone # _____
                                      Home Phone # _____

Alternate _____ Phone # _____
                                 Home Phone # _____

Accounting _____ Phone # _____
                                      Home Phone # _____

Alternate _____ Phone # _____
                                 Home Phone # _____

Legal Counsel _____ Phone # _____
                                         Home Phone # _____

Alternate _____ Phone # _____
                                 Home Phone # _____

Public Relations _____ Phone # _____
                                            Home Phone # _____

Alternate _____ Phone # _____
                                 Home Phone # _____

Technical _____ Phone # _____
                                     Home Phone # _____

Alternate _____ Phone # _____
                                 Home Phone # _____

Sales _____ Phone # _____
                                 Home Phone # _____

Alternate _____ Phone # _____
                                 Home Phone # _____

Marketing _____ Phone # _____
                                     Home Phone # _____

Alternate _____ Phone # _____
                                 Home Phone # _____

Consumer Affairs _____ Phone # _____
                                            Home Phone # _____

Alternate _____ Phone # _____
                                 Home Phone # _____

Other _____ Phone # _____
                                 Home Phone # _____

_____ Phone # _____
                       Home Phone # _____

_____ Phone # _____
                       Home Phone # _____

Brokers _____ Phone # _____
                                   Home Phone # _____

_____ Phone # _____
                       Home Phone # _____

_____ Phone # _____
                       Home Phone # _____

THE RECALL COORDINATOR

One person should be identified as the Recall Coordinator to prepare for and coordinate all activities related to recalls or withdrawals. The Recall Coordinator should be knowledgeable about every aspect of the company's operations, including purchasing, processing, quality assurance, distribution and consumer complaints.

The Recall Coordinator should select the other members of the Recall Team. Each member of the Team should have a list of office, home and weekend telephone numbers for every other member of the Recall Team. The Coordinator should be authorized to make decisions on procedures to be used in carrying out a recall or withdrawal, and should report to top management at regular, specified intervals.

THE RECALL PLAN

The Recall Team should develop the company Recall Plan and keep it current. Once the basic structure and procedures are established, the Recall Team should meet periodically to see that its plan meshes with company policy and operating methods. The Recall Coordinator must have full authority to convene meetings of the Recall Team and other company personnel whenever there is a need, regardless of other activities that may be underway in the company.

Once the recall plan is developed, it should be practiced to make sure it works. The practice runs should test the company's ability to undertake a ready review of records related to processing, raw product, quality control and distribution for any product. A diagram of recall functions of various individuals and groups in an organization for a small company is shown in Figure 15-2.

DEFINING THE PROBLEM

The recall plan must include a method for verifying whether or not there is a problem and, if so, for defining its dimensions. How did the problem come to the company's attention? What is the reliability of the information that suggests there is a problem? If the problem is real, how serious is it and how extensive?

The recall plan must include methods of rapidly verifying:

- Any deviations in processing records, in container closure records, in cook check records; any breakdown or change in equipment or sanitation practices; or, any unusual consumer complaints for the product in question.
- Unusual occurrences in product preparation or handling not reflected in quality control or in processing records.
- The time period during which faulty product could have been produced.
- Affected product codes.
- Customers to whom the product has been shipped.
- The public health significance of the product defect.

Figure 15-2. Recall Organization Function Diagram

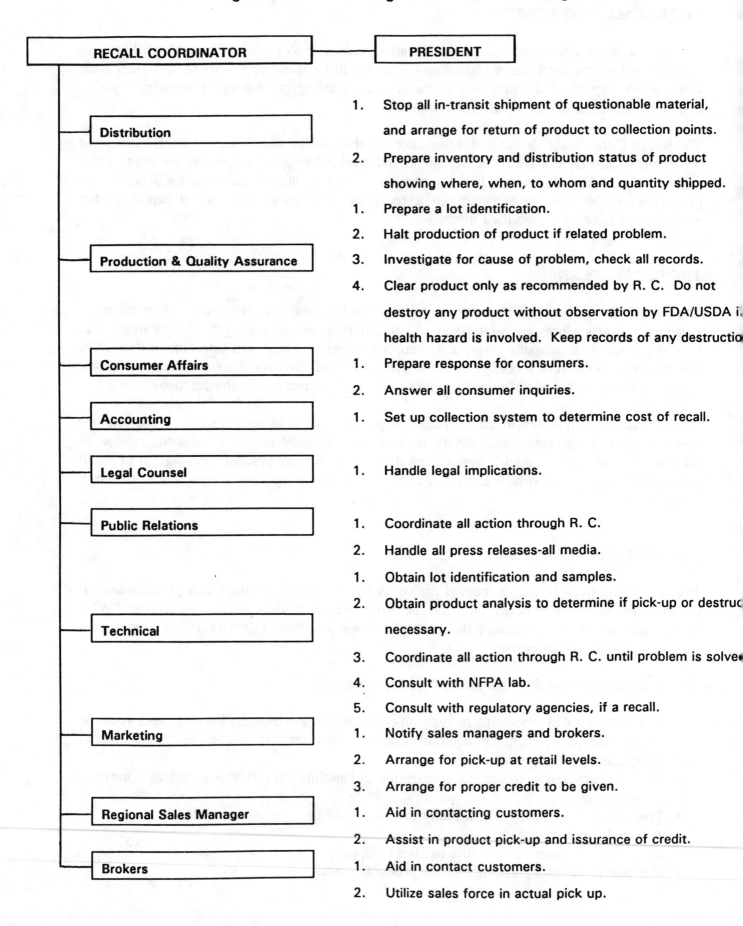

| RECALL COORDINATOR | ──▶ | PRESIDENT |

Distribution

1. Stop all in-transit shipment of questionable material, and arrange for return of product to collection points.
2. Prepare inventory and distribution status of product showing where, when, to whom and quantity shipped.

Production & Quality Assurance

1. Prepare a lot identification.
2. Halt production of product if related problem.
3. Investigate for cause of problem, check all records.
4. Clear product only as recommended by R. C. Do not destroy any product without observation by FDA/USDA i. health hazard is involved. Keep records of any destructio

Consumer Affairs

1. Prepare response for consumers.
2. Answer all consumer inquiries.

Accounting

1. Set up collection system to determine cost of recall.

Legal Counsel

1. Handle legal implications.

Public Relations

1. Coordinate all action through R. C.
2. Handle all press releases-all media.

Technical

1. Obtain lot identification and samples.
2. Obtain product analysis to determine if pick-up or destruc necessary.
3. Coordinate all action through R. C. until problem is solve
4. Consult with NFPA lab.
5. Consult with regulatory agencies, if a recall.

Marketing

1. Notify sales managers and brokers.
2. Arrange for pick-up at retail levels.
3. Arrange for proper credit to be given.

Regional Sales Manager

1. Aid in contacting customers.
2. Assist in product pick-up and issurance of credit.

Brokers

1. Aid in contact customers.
2. Utilize sales force in actual pick up.

OBTAINING INFORMATION

The Recall Team must get its facts first-hand. A written record must be kept showing the date, time of day, and details of discussions with everyone who is a source of needed information. A lack of facts and detail can waste time and effort and result in bad decision-making. There is a high probability that initial information, obtained early in the fact-gathering process, will be faulty.

Initial information indicating a possible need to withdraw product from distribution can come from a variety of sources, including:

- Consumer complaints
- Newspaper or radio or television reports
- State or federal regulatory officials
- Distributors or customers
- Sales representatives
- Ingredient suppliers
- Container suppliers
- Trade association contacts with regulatory officials
- Company quality assurance and production records

In the case of a reported illness, those closest to the situation should be contacted: the "victim" and family members, the attending physician, hospital authorities, local health officials, and other regulatory officials who may be involved. Records should contain the following information:

1. The name, address and phone number of the person or persons claiming an illness or injury and how the alleged occurrence was initially reported or discovered.
2. The type of alleged illness or injury with as many details as possible, including the name, address and phone number of the attending physician and his assessment of the case.
3. Complete identification of the product and its container, the involved code, the label name, and such facts as the condition of the container and its handling and storage by the consumer.
4. Store where the product was purchased, including address, telephone number and store personnel involved or aware of the incident.
5. Location of the complaint sample and an assessment of whether or not it can be obtained for testing.
6. The names, titles, addresses and phone numbers of any health authorities and law enforcement officials who are involved or may become involved.
7. Results of any laboratory testing of samples from the suspect products.
8. Any consumer contact with legal counsel, including the name, address and telephone number of the attorney, if available.
9. Any action by the store or retail chain to remove product from store shelves, and who authorized and carried out the removal.

COMMUNICATING DURING A RECALL

Clear, concise and accurate communications during a product emergency is critical. The recall plan should spell out who in the company is to be notified of developments at various stages of the recall or withdrawal process. The plan should also state which outside parties--regulatory agencies, associations, distributors, customers, and news media--are to be notified, and at what stages of the process. A model recall coordination outline is shown in Figure 15-3.

One of the first steps in a recall is to notify all brokers, distributors and retailers who have received the suspect product that it shall be withdrawn from distribution and held pending further notice from the manufacturer. This notification should include the reason for the recall.
Furthermore, the recall or withdrawal communication should be in the form of a telegram, fax, mailgram or first class letter. If a letter is used, the letter and the envelope should be marked "Urgent."

PHYSICAL RECOVERY OF THE PRODUCT

The plan must include detailed instructions for achieving removal from distribution channels. Marketing and sales people are often key players in getting the product out of the marketplace. They must be able to identify customers and shipping destinations and to pinpoint the location and quantity of suspect product.

Unless some provision for paying for the product or replacing with similar but safe product is provided by the food manufacturer, distributors are less likely to isolate and protect each container of the recalled product. Counts on the recalled containers are very important. It is also important that no containers other than those of the recalled codes are returned to the manufacturer.

The recalled product should be held in a separate warehouse or isolated by lock and key from other (good) product. The final determination of the disposition of the recalled product must be done in coordination with FDA or USDA.

RECORDS

The importance of adequate records detailing production and distribution of products cannot be stressed enough. In recovering product, it is helpful if cases bear the same codes as the packages inside them. Only one code should be packed in each case. Codes should be changed frequently, so that lots remain small, and pick-ups from distribution centers or stores can be targeted to include only suspect merchandise. Processors who pack for private label distribution should have detailed records on labels used for product shipped to each customer.

All physical evidence and records related to product complaints of a critical nature must be handled in a way that demonstrates that your company and its employees were not negligent in handling, storing and testing evidence or samples.

Figure 15-3. Recall Action Plan Diagram

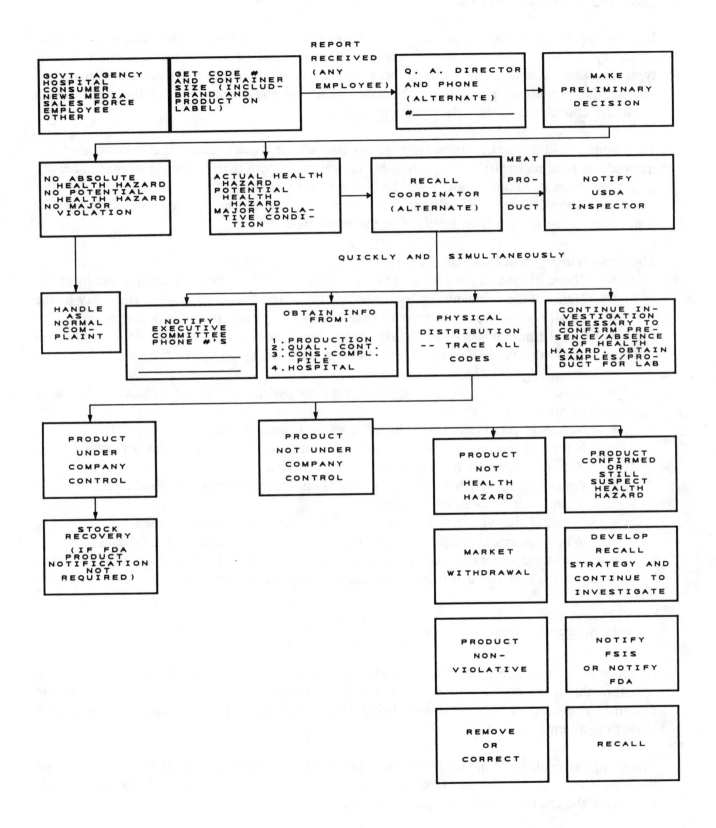

ISSUING A PRESS RELEASE

In a product recall, FDA or FSIS may insist upon issuance of a news release if it is considered necessary to alert consumers to the existence of a potential hazard to health. The company typically resists issuing a press release because the adverse publicity it creates can damage the company's reputation, generate an increase in unwarranted consumer claims, and lessen consumer confidence in the safety of its products.

The company should argue strenuously against issuance of a news release if it is convinced that its products pose no hazard to public health. However, if there is an imminent danger to health, long-term interests of the company will be best served by agreeing to issue a news release. The recall plan should have a "model" press release or outline that can be used to quickly prepare a statement covering the facts involved in the immediate case.

The press release should contain the following basic information:
- Name of the company making the withdrawal and its corporate headquarters location.
- Name of the company spokesperson and his or her telephone number (this may be the recall coordinator or someone in the company's public relations department).
- Label name or names of the affected product, codes and container sizes involved in the withdrawal, limited to only those names, codes and sizes for which there is a potential public health hazard.
- Area of distribution, naming each state to which the suspect product codes were shipped.
- Names of retail outlets to which the suspect product codes were shipped.
- Nature of the problem or potential problem with the product codes involved.
- Risk to consumers of the product codes involved.
- Period during which the suspect product codes were manufactured and distributed and the number of cases or containers involved.
- Status of the withdrawal action on the day the news release is issued.
- What consumers should do with the product if they have it in their possession.
- A statement that the company is cooperating with the appropriate regulatory officials in its action and that it has corrected the cause of the problem.

THE SPOKESPERSON

A company engaged in a product withdrawal should designate a single spokesperson to represent it before the news media. Even if the company does not issue a press release, it should designate a spokesperson, because work being done to recover the product from distribution may attract the attention of the media.

The company spokesperson should be someone who is believable and credible. In discussions with reporters, he or she must have the ability to state facts openly and honestly, while protecting the interests of the company and consumers alike.

Everyone in the company should know who the spokesperson is, and all media inquiries should be directed to the spokesperson. No one else should talk to the media about the recall.

It is important that the spokesperson communicate how much the company cares about the safety of its products. It is important for the media to know that the company either has corrected the problem, or is taking all steps possible to find and correct the problem.

It is also important that the company limit the scope of the problem to as few codes, time periods and geographical areas as possible. Defining the problem in such a way helps reporters and consumers understand that there will be a point beyond which there need be no concern.

Here are some basic steps to follow when responding to news media questions:

1. Avoid saying "no comment." It creates the impression you are hiding something.
2. Have a prepared statement you can give to reporters.
3. Don't be too technical. Try to stick to the facts in the prepared release.
4. Be sure to make the basic points in your prepared statement, even if reporters don't ask questions that lead naturally into those points.
5. Think about your answer before you reply. If you don't known the answer to a question, tell the reporter you don't know but will find out and get back to him or her as soon as possible.
6. Try to help reporters meet their deadlines.
7. Be honest. Don't respond to speculation. Don't place blame.

RECALL COORDINATION OUTLINE

1. Institute a log showing date and time of events and actions taken.
2. Correlate and evaluate known information on the extent and nature of the problem.
3. Discontinue distribution of affected product code(s).
4. Notify (in this order): company officials; trade associations; regulatory agency, if required; and suppliers, as necessary, .
5. Decide whether to recall or to make further investigation.
6. Determine area and level of distribution of product code(s).
7. Continue investigation of nature, extent, cause and remedy of problem as necessary. Look at processing, packaging, warehouse spoilage, consumer claims records, and cook checks, if available.
8. Re-examine recall decision, if necessary.
9. Notify and instruct company staff.
10. Determine disposition of recalled product.
11. Prepare and issue, with regulatory agency concurrence, recall requests in form of letters, telegrams, faxes, and/or public statements to known distributors. Include necessary instructions.
12. Prepare and issue news releases as necessary, preferably with concurrence of regulatory agency.
13. Maintain records of recall progress.
14. Report recall progress to company officials and regulatory agency.
15. Complete effectiveness checks and final report.

KEY TERMS IN WITHDRAWALS AND RECALLS

The following are definitions for some of the key terms used in product withdrawal and recall situations:

Action Level: The level of contamination informally established by FDA and USDA at which a product becomes subject to regulatory action. Action levels or defect action levels (sometimes called DALs) are set for contaminants such as pesticides, insects, and filth that cannot be avoided in the manufacturing process.

Adulteration: A product is adulterated when it contains a poisonous or deleterious substance, an unauthorized food additive, an unavoidable contaminant or a pesticide residue in excess of established action or tolerance levels. FDA action levels for added poisonous or deleterious substances in food are not binding. The agency may institute enforcement activity at levels below the action level or elect *not* to take action when levels exceed the established action level.

Consumer Alert: A news release provided to wire services, major newspapers and possibly to radio or television stations to inform the public that a particular food product may present a hazard to health.

Correction: Repair, modification, adjustment, relabeling, destruction or inspection of a product to cause it to conform to applicable regulations.

Effectiveness Checks: Follow-up by the company for FDA or FSIS through visits, phone calls, letters and other means to assure that all distributors and purchasers of a product undergoing withdrawal or recall have been notified and are returning the product.

Firm-Initiated Recall: A product correction or recall undertaken by the food manufacturer on its own initiative.

Government-Initiated Recall: A recall undertaken by a food manufacturer in response to a request from FDA or FSIS.

Health Hazard Evaluation: Unofficially, an evaluation of the health hazard(s) posed by a product being considered for withdrawal or recall, conducted by the manufacturer or others. Officially, such an evaluation conducted by FDA or FSIS.

Injunction: A court order obtained by a federal or state agency to prevent a food manufacturer or retailer from carrying out all or part of its business operations.

Market Withdrawal: Removal from distribution of substandard products by the manufacturer, usually without publicity.

Misbranding: Failure of a food product to comply with applicable labeling or composition requirements.

Recall Classification: The ranking of a recall as Class I, II or III, by FDA or FSIS, based upon the degree to which the product being recalled poses a hazard to health.

Recall Plan: The overall plan developed by each food manufacturer setting forth procedures to be followed in the event a product must be removed from distribution channels.

Recall Strategy: The course of action followed in conducting the recall of a specific product.

Seizure: Legal action in the courts by a federal or state agency to gain legal control over a product not in compliance with applicable regulations controlling safety, wholesomeness, composition or labeling.

REFERENCES

FDA. 1988. Chapter 5-00. Recall procedures. In, *Regulatory Procedures Manual*. Food and Drug Administration, Washington, D.C.

FSIS. 1992. *Recall of Inspected Meat and Poultry Products*. FSIS Directive 8080.1, Revision 2, November 3, 1992. USDA/FSIS, Washington, D.C.

Health Protection Branch. 1984. *Product Recall Procedures*. Health Protection Branch, Health and Welfare Canada, Ottawa, Ontario.

NFPA. 1987. *NFPA Bulletin 34-L: Successfully Managing Product Recalls & Withdrawals*. National Food Processors Association, Washington, D.C.

NFPA. 1988. *Manual on Pre-Emergency Planning and Disaster Recovery*. National Food Processors Association, Washington, D.C.

SECTION VI

APPENDIX

Contents of HACCP/QC Manual

CONTENTS OF HACCP/QC MANUAL

INTRODUCTION

The following pages in Appendix I present an example of a detailed Table of Contents for a HACCP/QC Manual for the production of thermally processed (shelf-stable) beef stew in cans. This example is intended to represent a comprehensive manual which includes product quality, product safety and regulatory issues.

This type of manual is commonly used by small to medium-sized companies which have HACCP Plans. Although large companies can benefit from the complete separation of the HACCP Plan from quality and regulatory issues, smaller companies usually employ programs which partially combine these programs.

Many supervisors of operations feel more comfortable in dealing with the combination of safety and non-safety issues, since they believe this allows a better chance to oversee all of the programs. However, it is important to separate out the safety issues in a distinct HACCP Program, such as that presented in the example. For instance, while the Master Diagram (page 3.4) would include CCPs, QC control points and regulatory control points, the each of these three areas would be presented on separate flow diagrams and in separate descriptions detailing the (separate) CCPs and control points.

Note that this is not a small manual, because it contains a large number of reference standards, regulations and specifications. While some individuals would hesitate to include all of this documentation in a single manual, it is necessary to include all of this information so the document control system will have a chance to function effectively.

HACCP/Q.C. MANUAL

TABLE OF CONTENTS

TABLE OF CONTENTS (Continued)